# THE NOTRE DAME LECTURES

# LECTURE NOTES IN LOGIC

## A Publication of
## THE ASSOCIATION FOR SYMBOLIC LOGIC

LECTURE NOTES IN LOGIC    18

# THE NOTRE DAME LECTURES

Edited by

Peter Cholak
Department of Mathematics
University of Notre Dame

ASSOCIATION FOR SYMBOLIC LOGIC

CRC Press
Taylor & Francis Group
Boca Raton  London  New York

CRC Press is an imprint of the
Taylor & Francis Group, an **informa** business

AN A K PETERS BOOK

CRC Press
Taylor & Francis Group
6000 Broken Sound Parkway NW, Suite 300
Boca Raton, FL 33487-2742

First issued in hardback 2020

ISBN-13: 978-1-56881-249-6 (hbk)
ISBN-13: 978-1-56881-250-2 (pbk)

Visit the Taylor & Francis Web site at
http://www.taylorandfrancis.com

and the CRC Press Web site at
http://www.crcpress.com

**Library of Congress Cataloging-in-Publication Data**

The Notre Dame lectures / edited by Peter Cholak.
    p. cm. – (Lecture notes in logic ; 18)
   Includes bibliographical references and index.
   ISBN 1-56881-249-3 (acid-free paper) – ISBN 1-56881-250-7 (pbk. : acid-free paper)
   1. Logic, Symbolic and mathematical. I. Cholak, Peter, 1962- II. Series
QA9.2.N67 2005
511.3–dc22                                                       2004065969

Publisher's note: This book was typeset in LaTeX, by the ASL Typesetting Office, from electronic files produced by the authors, using the ASL documentclass asl.cls. The fonts are Monotype Times Roman. The cover design is by Richard Hannus, Hannus Design Associates, Boston, Massachusetts.

# PREFACE

In the fall of 2000, the Notre Dame logic community was lucky enough to get Greg Hjorth, Rodney G. Downey, Zoé Chatzidakis, and Paola D'Aquino to come to Notre Dame. Each of them came for about a month and taught a month long graduate class on a topic of their choice. The following (refereed) articles are refinements of these lectures.

These classes were attended by Professors Steve Buechler, Peter Cholak, Julia Knight and Sergei Starchenko. The graduate students attending were Bijan Afshordel (visiting from Germany), Andrew Arana, Alexander Berenstein, Tom Doherty, Jacob Heidenreich, Evgueni Vassiliev, and Rebecca Weber.

We would like to thank the Notre Dame Department of Mathematics for allowing the logicians at Notre Dame to split one of the advanced graduate logic classes into four separate pieces and for a portion of the required funding. We would like to thank the Graduate School at Notre Dame for another portion of the funding. Many thanks to Julia Knight, the Charles L. Huisking Professor of Mathematics, Notre Dame, who used her chair account for the remaining funding.

Most of all we want to thank Greg Hjorth, Rodney G. Downey, Zoé Chatzidakis, and Paola D'Aquino for coming to Notre Dame and providing us with a very interesting series of lectures, and for the time and energy needed to turn them into the following articles. Our thanks.

Peter Cholak, Editor
for the Notre Dame logic community

# TABLE OF CONTENTS

# COUNTABLE MODELS AND THE THEORY
# OF BOREL EQUIVALENCE RELATIONS

## CONTENTS

**§1. Overview.** The main goal of these lecture notes will be to survey some of the recent work on Borel and $\underset{\sim}{\Sigma}^1_1$ equivalence relations. We will specifically consider the $\underset{\sim}{\Sigma}^1_1$ equivalence relation which arises as the isomorphism relation on a class of countable structures.

DEFINITION 1.1. A topological space is *Polish* if it is separable[1] and it allows a complete metric.

EXAMPLES 1.2.
1. $\mathbb{R}, \mathbb{C}$ in their usual topologies.
2. Any countable ordinal in the order topology where the basic open sets are $\{\beta : \gamma_1 < \beta < \gamma_2\}$. (This requires some proof of course. One way to do this is to show by transfinite induction on $\alpha < \omega_1$ that for any $a < b \in \mathbb{R} \cup \{+\infty\}$ we can find a closed homeomorphic copy of $\alpha$ in $[a, b)$ if $\alpha$ is a limit ordinal or in $[a, b]$ if $\alpha$ is a successor.)

---

The author gratefully acknowledges improvements suggested by Wesley Calvert, Peter Cholak, Jacob Heidenreich, and the other logicians at Notre Dame who commented on the talks on which these notes are based.

[1]I.e., there is a countable dense subset.

**The Notre Dame Lectures**
Edited by P. Cholak
Lecture Notes in Logic, 18
© 2005, ASSOCIATION FOR SYMBOLIC LOGIC

3. $\mathbb{N}^{\mathbb{N}} = \{f : \mathbb{N} \to \mathbb{N}\}$. Here we can take as our complete metric

$$d(f, g) = 2^{-\Delta(f,g)}$$

where $\Delta(f, g)$ is the least $n$ at which $f(n) \neq g(n)$.

Actually for us some of the most important examples of Polish spaces will arise from model theory.

4. Let $\mathcal{L}$ be some countable language and let $\text{Mod}(\mathcal{L})$ be the space of $\mathcal{L}$-structures on $\mathbb{N}$ with the topology whose basic open sets are all those of the form

$$\{\mathcal{M} \mid \mathcal{M} \models \varphi(n_1, n_2, \ldots, n_k)\}$$

for $k, n_1, \ldots, n_k \in \mathbb{N}$ and $\varphi$ a quantifier-free formula.

So as a simple case of this let us say that $\mathcal{L} = \{U, R\}$ where $U$ is a unary function and $R$ is a binary relation. Then we can associate to any $\mathcal{M} \in \text{Mod}(\mathcal{L})$ the pair

$$\left(U^{\mathcal{M}}, \chi_{(R^{\mathcal{M}})}\right) \in \mathbb{N}^{\mathbb{N}} \times \{0, 1\}^{\mathbb{N}},$$

where $U^{\mathcal{M}}$ is $\mathcal{M}$'s interpretation of the unary function symbol and $\chi_{(R^{\mathcal{M}})}$ is the characteristic function of the binary relation from the point of view of $\mathcal{M}$.

We obtain as in Example 3 that $\mathbb{N}^{\mathbb{N}} \times \{0, 1\}^{\mathbb{N}}$ is Polish.[2] And our association

$$\mathcal{M} \mapsto \left(U^{\mathcal{M}}, \chi_{(R^{\mathcal{M}})}\right)$$

is a homeomorphism.

5. Actually, as we will see later, the topology in which we take as basic open sets those of the form

$$\{\mathcal{M} \mid \mathcal{M} \models \varphi(n_1, n_2, \ldots, n_k)\}$$

$\varphi$ a *first-order formula* gives rise to a Polish topology, richer than the first, but obviously closely connected.

6. More generally, if $F \subset \mathcal{L}_{\omega_1,\omega}$ is any countable "fragment"[3] then the topology generated by sets of the form

$$\{\mathcal{M} \mid \mathcal{M} \models \varphi(n_1, n_2, \ldots, n_k)\}$$

for $\varphi \in F$ is Polish.

DEFINITION 1.3. A subset of a Polish space is *Borel* if it can be generated from the open sets by the operations of complementation, countable union, and countable intersection. A function

$$f : X \to Y$$

is *Borel* if $f^{-1}[B]$ is Borel for any Borel $B \subset Y$. (Equivalently we simply have required that the pullbacks of open sets along $f$ are all Borel; the point is

---

[2] For instance, $d((f_1, f_2), (g_1, g_2)) = 2^{-\Delta(f_1,g_1)} + 2^{-\Delta(f_2,g_2)}$.

[3] Again, the definition of $\mathcal{L}_{\omega_1,\omega}$ (Defn 4.1) and "fragment" (Defn 4.3) will be given in Section 4.

that the operations of complementation, intersection, and union behave nicely with respect to pullback.)

An equivalence relation $E$ on a Polish space $X$ is *Borel* if it is Borel as a subset of $X \times X$ in the product topology.

With respect to this last definition, it is not hard to see that the product of two Polish spaces is Polish; in particular $X \times X$ is Polish.

Here a theorem dating back to the early 1900s describes the possible structure of an arbitrary Borel set.

THEOREM 1.1 (Classical). *If $B$ is a Borel set then either*

(i) $|B| \leq \aleph_0$ ($B$ *is countable*), *or*
(ii) $|B| = 2^{\aleph_0}$ ($B$ *has cardinality continuum*).

*Moreover, any two Borel sets of the same cardinality are Borel isomorphic.*

By this last part of the theorem I mean the following. If $X_1, X_2$ are Polish and $B_i \subset X_i$ ($i = 1, 2$) are Borel with $|B_1| = |B_2|$ then there is a function

$$f : B_1 \to B_2$$

with

(a) $f$ a bijection;
(b) $f^{-1}[C]$ Borel all $C \subset B_2$ Borel; and
(c) $f[C]$ Borel all $C \subset B_1$ Borel.

(Just in passing I should add that (c) above follows from the conjunction of (a) and (b)—but we probably won't need that particular fact.)

The situation for Borel equivalence relations initially appears the same.

THEOREM 1.2 (Silver). *Let $E$ be a Borel equivalence relation on a Polish space $X$. Then either*

(i) $|X/E| \leq \aleph_0$ ($E$ *has only countably many equivalence classes*), *or*
(ii) $|X/E| = 2^{\aleph_0}$ ($E$ *has continuum many classes*).

But there is no "moreover". The problem is with part (ii) where there is an enormous range of possible types of equivalence relations with continuum many equivalence classes.

DEFINITION 1.4. Let $E$ and $F$ be Borel equivalence relations on Polish $X$ and $Y$. We write

$$E \leq_B F$$

and say that $E$ is *Borel reducible* to $F$ if there is a Borel function

$$f : X \to Y$$

such that, for all $x_1, x_2 \in X$,

$$x_1 E x_2 \Leftrightarrow f(x_1) F f(x_2).$$

$E \sim_B F$ indicates that we have both

$$E \leq_B F$$

and

$$F \leq_B E.$$

EXAMPLES 1.5.

1. For any Polish space $X$ we have $\mathrm{id}(X)$ as the identity relation on $X$: $\{(x_1, x_2) \in X \times X : x_1 = x_2\}$.

2. $E_0$, eventual agreement on

$$2^{\mathbb{N}} = \{0, 1\}^{\mathbb{N}},$$

the space of infinite binary sequences. So that $f_1 E_0 f_2$ if there is some $N \in \mathbb{N}$ such that

$$\forall m > N \big( f_1(m) = f_2(m) \big).$$

3. $E_v$, the Vitali equivalence relation consisting of cosets of $\mathbb{Q}$ in $\mathbb{R}$. So $r_1 E_v r_2$ if

$$r_1 - r_2 \in \mathbb{Q}.$$

Actually it turns out that $E_0 \sim_B E_v$ and for any two uncountable Polish spaces $X_1, X_2$, one has $\mathrm{id}(X_1) \sim_B \mathrm{id}(X_2)$.

Possibility (ii) from Silver's theorem represents the start of a very long story. The first major result was obtained at the end of the 1980s:

THEOREM 1.3 (Harrington-Kechris-Louveau). *Let $E$ be a Borel equivalence relation. Then exactly one of the following holds*:

(i) $E \leq_B \mathrm{id}(2^{\mathbb{N}})$;

(ii) $E_0 \leq_B E$.

It takes a nontrivial amount of descriptive set theory to prove this theorem. Instead I will probably try to work through the analogue for isomorphism relation on countable structures.

Just as a warning, for $\mathcal{L}$ a countable language we do not have that the isomorphism relation on countable $\mathcal{L}$-structures is Borel in any of the previously mentioned Polish topologies, except when $\mathcal{L}$ is composed solely of unary predicates. Thus Harrington-Kechris-Louveau will not generally apply.

However, it turns out that an analogue can be proved.

Here for $\sigma \in \mathcal{L}_{\omega_1, \omega}$ I will let $\mathrm{Mod}(\sigma)$ be the subset of $\mathrm{Mod}(\mathcal{L})$ consisting of models of $\sigma$. This is a Borel subset of $\mathrm{Mod}(\mathcal{L})$ under any of our topologies, and a Polish space in its right in the topology generated by any countable fragment $F \subset \mathcal{L}_{\omega_1, \omega}$ which includes $\sigma$.

THEOREM 1.4 (Becker; Hjorth-Kechris). *For $\mathcal{L}$ a countable language and $\sigma \in \mathcal{L}_{\omega_1,\omega}$ we have either*

(i) *there is a $\underset{\sim}{\Delta}_1^{HC}$ function[4] assigning elements of $2^{<\omega_1}$ as complete invariants,* or

(ii) $E_0 \leq_B \cong |_{\mathrm{Mod}(\sigma)}$.

Beyond $E_0$ the picture becomes extremely complicated and it is poorly understood. Another side topic will be the theory of continuous actions of Polish groups and some special cases of the *topological Vaught conjecture*; this should fit naturally into the proof of "Becker; Hjorth-Kechris" above, since it too requires some discussion of the general actions of Polish groups.

## §2. Borel sets.

DEFINITION 2.1. A topological space is *Polish* if it is separable and there exists a complete metric which generates its topology.

Note then that a Polish space always has a countable basis for its topology.

Just to give some context, let me mention a classical characterization of Polish spaces. We will never use this result, so I mention it without proof.

THEOREM 2.1 (Hausdorff). *A topological space is Polish if and only if it is homeomorphic to some subset of the Hilbert cube, $[0,1]^{\mathbb{N}}$, and it is the image of $\mathbb{N}^{\mathbb{N}}$ under some continuous, open function.*

DEFINITION 2.2. The $\sigma$-algebra generated by the open sets forms the *Borel* subsets of a Polish space.

EXAMPLE 2.3. Let $\mathcal{L}$ be a countable language and $\mathrm{Mod}(\mathcal{L})$ the space of $\mathcal{L}$-structures on $\mathbb{N}$; if we equip this space with the topology generated by quantifier-free formulas then it is naturally homeomorphic to

$$\prod_{\substack{\text{no. of unary predicates}}} \{0,1\}^{\mathbb{N}} \times \prod_{\substack{\text{no. of binary relations}}} \{0,1\}^{\mathbb{N}\times\mathbb{N}}$$

$$\times \prod_{\substack{\text{no. of ternary relations}}} \{0,1\}^{\mathbb{N}\times\mathbb{N}\times\mathbb{N}}$$

$$\times \cdots \prod_{\substack{\text{no. of unary functions}}} \mathbb{N}^{\mathbb{N}} \times \prod_{\substack{\text{no. of binary function}}} \mathbb{N}^{\mathbb{N}\times\mathbb{N}} \cdots .$$

We will shortly prove that countable products of Polish spaces are Polish, and so this is as well.

---

[4]That is, some function which is $\Delta_1(x)$ over the hereditarily countable sets for some hereditarily countable $x$. The details of this definition are not so important; it is just that we can not use the old standby of Borel reduction for various technical reasons.

One might choose to think of the complexity class $\underset{\sim}{\Delta}_1^{HC}$ as being the analogue to HC of recursive to HF (the hereditarily finite sets).

Then one has for any first-order formula $\varphi$ and $\vec{n} \in \mathbb{N}^{<\mathbb{N}}$ that the set of

$$\{ \mathcal{M} \mid \mathcal{M} \models \varphi(\vec{n}) \}$$

is a Borel set. This is easily proved by induction on the logical complexity of $\varphi$, starting with the quantifier-free formulas as the base case, and working up through the addition of existential and universal quantifiers[5] by observing that

$$\{ \mathcal{M} : \mathcal{M} \models \exists x \varphi(\vec{n}, x) \} = \bigcup_{m \in \mathbb{N}} \{ \mathcal{M} : \mathcal{M} \models \varphi(\vec{n}, m) \}$$

and

$$\{ \mathcal{M} : \mathcal{M} \models \forall x \varphi(\vec{n}, x) \} = \bigcap_{m \in \mathbb{N}} \{ \mathcal{M} : \mathcal{M} \models \varphi(\vec{n}, m) \} .$$

Thus the topology of quantifier-free formulas and the topology of first-order logic give rise to the same Borel sets.

DEFINITION 2.4. A set $X$ with a $\sigma$-algebra of sets $\mathcal{B}$ is said to be a *standard Borel space* if there is some Polish topology which gives rise to $\mathcal{B}$ as its Borel sets.

Now we can at least state the temporary goals, just for this section.
1. Mod($\mathcal{L}$) in the topology of first-order formulas is Polish.
2. A Borel subset of a standard Borel space is again standard Borel.
3. Any uncountable Borel set has size $2^{\aleph_0}$.

Actually in the proof of this last point we will follow the path chosen in Kechris's *Classical descriptive set theory*, reducing it to (2) and the perfect set theorem for Polish spaces.

The arguments here are all fairly elementary. One does little calculations with respect to metrics, and the typical step is to show that we can fiddle some coordinates to produce a complete metric with various desirable properties.

Ultimately we will use these little results as part of the proof of the above-mentioned dichotomy theorem for the isomorphism relation on classes of countable structures. The proof requires us to repeatedly modify a given space of structures, through transfinitely many steps, and at each stage it will be necessary to maintain the Polishness of the space.

The study of Polish space has emerged as a central theme in descriptive set theory. It is generally accepted that these are the spaces we can work with, applicable to a wide context but sufficiently specific to avoid certain pathologies.

Recently a few people—notably Becker and Kechris—have made some use of the more general "strong Choquet" spaces. But it is not clear that this is likely to ever unseat the current prejudice in favor of Polish.

---

[5]Since every formula is provably equivalent to one with all the quantifiers on the outside and all the Boolean connectives on the inside, this is all we need to worry about.

LEMMA 2.5. *Let $X$ be a Polish space. Then it has a compatible complete metric bounded by 1.*

PROOF. Let $d'$ be some complete metric. Let

$$d(x_1, x_2) = \frac{d'(x_1, x_2)}{1 + d'(x_1, x_2)}.$$

The transformation

$$r \mapsto \frac{r}{1 + r}$$

provides an order preserving transformation of $[0, \infty)$ onto $[0, 1)$. Thus, jumping ahead to the end of the argument, once we establish that $d$ *is* a metric then we have the same Cauchy sequences for the two metrics, and from this, completeness follows for free. Similarly the two metrics have the same open balls, and hence give rise to the same topologies.

We are therefore only left with proving it to be a metric. And here the main issue is showing that the triangle inequality holds.

Let $x_1, x_2, x_3 \in X$. We want to show

$$d(x_1, x_3) \leq d(x_1, x_2) + d(x_2, x_3).$$

We may assume that

$$d'(x_1, x_3) \geq d'(x_1, x_2), d'(x_2, x_3);$$

since if for instance $d'(x_1, x_3) \leq d'(x_1, x_2)$ then

$$d(x_1, x_3) \leq d(x_1, x_2),$$

since our change in coordinates is order preserving, and we are done.

But now we have

$$d(x_1, x_2) + d(x_2, x_3) = \frac{d'(x_1, x_2)}{1 + d'(x_1, x_2)} + \frac{d'(x_2, x_3)}{1 + d'(x_2, x_3)}$$

$$\geq \frac{d'(x_1, x_2) + d'(x_2, x_3)}{1 + d'(x_1, x_3)},$$

by our assumption above, which in turn, by the triangle inequality for $d'$, is at least as large as $\frac{d'(x_1, x_3)}{1 + d'(x_1, x_3)}$. ⊣

COROLLARY 2.6. *Let $(X_i)_i$ be a sequence of Polish spaces. Then*

$$\prod_{i \in \mathbb{N}} X_i$$

*is Polish.*

PROOF. Let $d_i$ be a complete metric on $X_i$, $i \in \mathbb{N}$. By Lemma 2.5 we may assume the metric bounded by one. Then for

$$\vec{x} = (x_0, x_1, x_2, \dots), \vec{y} = (y_0, y_1, y_2, \dots) \in \prod_{i \in \mathbb{N}} X_i$$

we can let

$$d(\vec{x}, \vec{y}) = \sum_{i \in \mathbb{N}} 2^{-i} d_i(x_i, y_i).$$

⊣

LEMMA 2.7. *An open subset of a Polish space is Polish in the subspace topology.*

PROOF. Let $X$ be a Polish space, $d$ a complete metric, $O \subset X$ open.

Let us fix a continuous

$$\pi: O \to \mathbb{R}$$

which has the property

$$\pi(x) \to +\infty$$

as $x \to X \setminus O$. For instance we can take

$$\pi(x) = \frac{1}{\inf\{d(x, y) \mid y \notin O\}}.$$

Then let

$$d_O(x_1, x_2) = d(x_1, x_2) + |\pi(x_1) + \pi(x_2)|.$$

Since $\pi$ is continuous we do not add to the topology of $O$ by taking this metric. For instance, if $V \subset O$ is open, $x \in V$, then we can find some   such that for all $y$ with $d(x, y) <$   we have $y \in V$; but then from the definition of $d_O$ we have every $y$ with $d_O(x, y) <$   in $V$, which is just what we need to show $V$ is open with respect to $d_O$. Conversely if $V \subset O$ is $d_O$-open, $x \in V$, then we can choose some   with the open ball of radius   with respect to $d_O$ included in $V$; and then the continuity of $\pi$ enables us to find some $\delta < \frac{1}{2}$ such that for all $y \in O$ with $d(x, y) < \delta$ we have

$$|\pi(x) - \pi(y)| < \frac{1}{2},$$

and thus the ball of radius $\delta$ with respect to $d$ is included in $V$.

And the various facts required for $d_O$ to be a metric are trivial to verify. The main issue is completeness.

But if $(x_i)_i$ is a Cauchy sequence in $(O, d_O)$ then it will necessarily be $d$-Cauchy, and so it will have a limit $x_\infty \in X$. We need to check $x_\infty \in O$.

But otherwise we obtain

$$\pi(x_i) \to +\infty$$

and hence for all $x_N$ on our sequence

$$|\pi(x_N) - \pi(x_i)| \to +\infty$$

as $i \to \infty$, contradicting our assumption on the sequence $(x_i)_i$. ⊣

DEFINITION 2.8. A subset of a Polish space is $G_\delta$ if it is given by a countable intersection of open sets.

Some people use $\underset{\sim}{\Pi^0_2}$ instead of $G_\delta$. And in fact this is a better notation, since it forms part of a general scheme to refer to the entire Borel hierarchy, while the

$$G_\delta, F_\sigma, G_{\delta\sigma}, F_{\sigma\delta}, \ldots$$

method only extends to the finite levels.

Note that every closed set is $G_\delta$. If $C \subset X$ is closed, then it equals

$$\bigcap_{n\in\mathbb{N}} \left\{ x : \exists y \in C \left( d(x, y) \le \frac{1}{n} \right) \right\}.$$

COROLLARY 2.9. *Any $G_\delta$ subset of a Polish space is Polish in the subspace topology.*

PROOF. Let $X$ be Polish,

$$B = \bigcap_{i\in\mathbb{N}} O_i$$

a $G_\delta$ subset, where each $O_i$ is open, and let $d_i$ be a complete metric on $O_i$ for $i \in \mathbb{N}$.

We may assume each $d_i$ is bounded by 1, and then let

$$d_B(x, y) = \sum_{i\in\mathbb{N}} 2^{-i} d_i(x, y).$$

$\dashv$

LEMMA 2.10. *For $\mathcal{L}$ a countable language, the space $Mod(\mathcal{L})$ equipped with the topology of first-order formulas is a Polish space.*

PROOF. Let $\{c_n \mid n \in \mathbb{N}\}$ be fresh constants and let $\hat{\mathcal{L}} = \mathcal{L} \cup \{c_n \mid n \in \mathbb{N}\}$ be the language obtained by adding them to $\mathcal{L}$. Let $S$ be the collection of sentences in $\hat{\mathcal{L}}$. Then

$$\{0, 1\}^S$$

in the product topology is naturally homeomorphic to $\{0, 1\}^{\mathbb{N}}$, and thus Polish.

Let $B$ be the set of $f \in \{0, 1\}^S$ such that
(a) $\{\varphi \mid f(\varphi) = 1\}$ is consistent, and
(b) for all $\varphi$ we have

$$f(\varphi) = 0 \Leftrightarrow f(\neg\varphi) = 1,$$

and
(c) for all $\varphi$ we have

$$f(\exists x \varphi(x) = 1) \Leftrightarrow \exists n \in \mathbb{N}(f(\varphi(c_n)) = 1).$$

CLAIM. $B$ is a $G_\delta$ subset of $\{0, 1\}^S$.

PROOF OF CLAIM. Since countable intersections of $G_\delta$ sets are $G_\delta$, it suffices to show that the conditions (a), (b), and (c) all correspond to $G_\delta$ sets. This is almost trivially true for (a), since a first-order theory is inconsistent if and only if it includes a finite set of sentences which are inconsistent; thus if $\mathcal{I} \subset [S]^{<\mathbb{N}}$ is the set of finite inconsistent subset of $S$ then (a) corresponds to the $G_\delta$ set

$$\bigcap_{F \in \mathcal{I}} \bigcup_{\psi \in F} \{f \mid f(\psi) = 0\}.$$

Similarly (b) is $G_\delta$ (and as with (a), it is actually closed).

Finally, (c) corresponds to the conjunction of the sets

$$\bigcap_{\varphi \in S} \bigcap_{n \in \mathbb{N}} \{f \mid f(\varphi(c_n)) = 1 \Rightarrow f(\exists x \varphi(x)) = 1\}$$

and

$$\bigcap_{\varphi \in S} \left( \{f \mid f(\exists x \varphi(x)) = 0\} \cup \bigcup_{n \in \mathbb{N}} \{f \mid f(\varphi(c_n)) = 1\} \right),$$

and is thus $G_\delta$.                                                                          CLAIM ⊣

To each $\mathcal{M} \in \mathrm{Mod}(\mathcal{L})$ let $T_\mathcal{M} \in \{0,1\}^S$ be given by

$$T_\mathcal{M}(\varphi(c_{n_1}, c_{n_2}, \dots)) = 1 \Leftrightarrow \mathcal{M} \models \varphi(n_1, n_2, \dots).$$

It is more or less immediate from the definitions that each such $T_\mathcal{M}$ is in $B$.

CLAIM. The map

$$\mathrm{Mod}(\mathcal{L}) \to B$$
$$\mathcal{M} \mapsto T_\mathcal{M}$$

is a bijection of $\mathrm{Mod}(\mathcal{L})$ onto $B$.

PROOF OF CLAIM. The map is obviously one-to-one, since for any $\mathcal{M}_1 \neq \mathcal{M}_2$ there will be some atomic $\psi(\vec{x})$ and $n_1, n_2, \dots$ on which they disagree, whence $T_{\mathcal{M}_1}$ and $T_{\mathcal{M}_2}$ disagree on $\psi(c_{n_1}, c_{n_2}, \dots)$.

As for showing it onto, fix $T \in B$. We define $\mathcal{M}$ by the requirements that for any relation $R \in \mathcal{L}$

$$\mathcal{M} \models R(n_1, n_2, \dots)$$

if and only if $T(R(c_{n_1}, c_{n_2}, \dots)) = 1$, and for any function symbol $F \in \mathcal{L}$

$$F^\mathcal{M}(n_1, n_2, \dots) = \ell$$

if and only if $T(F(c_{n_1}, c_{n_2}, \dots) = c_\ell) = 1$ and $\ell$ is the least such natural number. (This *is* well defined. By (a) and (b) we have $T(\exists x(F(c_{n_1}, c_{n_2}, \dots) = x)) = 1$, and then by (c) we have some witness $c_\ell$.) Then an easy induction on logical complexity shows that for any first-order $\psi$

$$\mathcal{M} \models \psi(n_1, n_2, \dots) \Leftrightarrow T(\psi(c_{n_1}, c_{n_2}, \dots)) = 1.$$                                          CLAIM ⊣

But this basically finishes the lemma. It follows from the definitions of the topologies that

$$\mathcal{M} \mapsto T_{\mathcal{M}}$$

is a homeomorphism. Thus $\mathrm{Mod}(\mathcal{L})$ in the topology generated by first-order formulas is homeomorphic to $B$; and we know $B$ must be Polish since it is a $G_\delta$ subset of a Polish space. ⊣

Up to now I have done the usually sloppy, convenient, but not quite correct thing, and simply identified a topological space with its underlying set. Over the next couple of lemmas we need to be more precise. Thus let us agree to write $(X, \tau)$ for a Polish space with underlying set $X$ and topology $\tau$.

**LEMMA 2.11.** *Let $(X, \tau)$ be a Polish space and $\mathcal{O} \subset X$ open. Then there is a new topology Polish $\hat{\tau}$ such that*

(a) $\hat{\tau} \supset \tau$;
(b) *every $\hat{\tau}$-open set is $\tau$-Borel (and hence the two topologies give rise to the same standard Borel structure)*;
(c) $\mathcal{O}$ *is $\hat{\tau}$-clopen.*

**PROOF.** Let $d_X$ be a compatible complete metric on $(X, \tau)$. Since $\mathcal{O}$ is open we can find a complete metric $d_\mathcal{O}$ on $\mathcal{O}$ which is compatible with its subspace topology from $\tau$. By earlier lemmas we can assume both to be bounded by 1.

Now we can define $d$ on $X$ by cases.

(i) If $x, y \in \mathcal{O}$ then $d(x, y) = d_\mathcal{O}(x, y)$.
(ii) If $x, y \notin \mathcal{O}$ then $d(x, y) = d_X(x, y)$.
(iii) If $x \in \mathcal{O}, y \notin \mathcal{O}$ then $d(x, y) = 2$.
(iv) If $x \notin \mathcal{O}, y \in \mathcal{O}$ then $d(x, y) = 2$.

It is as if we have split the space up into two halves, $\mathcal{O}$ and $X \setminus \mathcal{O}$, each a distance of 2 away from the other. Any $d$-Cauchy sequence must eventually decide to stay in one of these two halves, and hence will be either $d_X$ or $d_\mathcal{O}$ Cauchy. ⊣

Really this proof is using the fact that the disjoint union of two Polish spaces is again Polish.

The next lemma appears to be a folklore result. I am going to give a proof I saw presented by Ramez Sami.

**LEMMA 2.12.** *Let $(X, \tau)$ be a Polish space and $B \subset X$ Borel. Then there is a richer Polish topology $\hat{\tau} \supset \tau$ such that*

(a) $\hat{\tau} \supset \tau$;
(b) *every $\hat{\tau}$-open set is $\tau$-Borel*;
(c) $B$ *is $\hat{\tau}$-clopen.*

**PROOF.** Let $\mathcal{C}$ be the collection of all $B \subset X$ for which we can find $\hat{\tau}$ satisfying the lemma. We have previously seen that this includes all open sets.

Then by the structure of the definition we have that it must be closed under complementation—that is, if $B \in C$ then $(X \setminus B) \in C$.

Therefore to show it forms a $\sigma$-algebra we need only show it is closed under countable intersections. For this the following observation suffices.

CLAIM. Suppose for each $i$ we have a Polish topology $\tau_i \supset \tau$. Then the union of these topologies generates a Polish topology.

PROOF OF CLAIM. Consider the space

$$\prod_{i \in \mathbb{N}} (X, \tau_i).$$

This is a product of Polish spaces, and hence Polish, as is the closed subset

$$\{\vec{x} \mid \forall i, j \in \mathbb{N}\, (x_i = x_j)\}$$

consisting of all constant sequences.

If we associate to each $x \in X$ the sequence with constant value $x$ then we obtain a homeomorphism of $X$ under the topology generated by $\bigcup_{i \in \mathbb{N}} \tau_i$ and $\{\vec{x} \mid \forall i, j \in \mathbb{N}(x_i = x_j)\}$ with the subspace topology.                CLAIM $\dashv$

$\dashv$

DEFINITION 2.13. A subset $P$ of a Polish space $X$ is *perfect* if it is
(a) nonempty,
(b) closed, and
(c) has no isolated points—that is, if $U \subset X$ is open with $|U \cap P| \geq 1$ then we actually have $|U \cap P| \geq 2$.

LEMMA 2.14. *If $X$ is an uncountable Polish space, then it contains a perfect set.*

PROOF. Let $\mathcal{B} = \{U_n \mid n \in \mathbb{N}\}$ be a countable basis for $X$. Then define $\mathcal{B}_0 \subset \mathcal{B}$ to be the set of $\{U_n \mid U_n \cap X \text{ is countable}\}$ and define $P$ to be the elements of $X$ not contained in any element of $\mathcal{B}_0$:

$$P = X \setminus \left( \bigcup \mathcal{B}_0 \right).$$

Since $\bigcup \mathcal{B}_0$ is a union of open sets, we certainly have that $P$ is closed.

If $P$ is empty, then $X$ is a countable union of countable sets and we have a contradiction.

So instead assume $P \neq \emptyset$, $x \in P$, and we try to show $x$ not isolated.

For this purpose fix open $U$ containing $x$. After possibly shrinking $U$ we can assume $U \in \mathcal{B}$. But then if $|U \cap P| = 1$ we would have

$$\left| U \setminus \left( \bigcup \mathcal{B}_0 \right) \right| = 1$$

and hence

$$|U| \leq \left| \bigcup \mathcal{B}_0 \right| + 1 \leq \aleph_0 + 1 = \aleph_0,$$

which would place $U$ into $\mathcal{B}_0$, contradicting $x \in P$.

It is implicit in this proof that any Polish space without isolated points must contain a perfect subset.

LEMMA 2.15. *If $P$ is a perfect subset of a Polish space, then $P = |2^{\aleph_0}|$, and in fact contains a homeomorphic copy of*

$$\{0,1\}^{\mathbb{N}}$$

*as a closed subset.*

PROOF. We first need an observation about the structure of open sets in our perfect set.

CLAIM. If $U$ is open with $U \cap P \neq \emptyset$ and $> 0$ then there are open $V_0, V_1$ with

1. $V_0 \cap P, V_1 \cap P \neq \emptyset$;
2. $\overline{V_0}, \overline{V_1} \subset U$;[6]
3. $V_0 \cap V_1 = \emptyset$;
4. $d(V_0), d(V_1) < .$[7]

PROOF OF CLAIM. Since $P$ has no isolated points we may choose $x_0, x_1 \in U \cap P$ with $x_0 \neq x_1$. Then we may choose sufficiently small open neighborhoods $B_{2\delta}(x_0), B_{2\delta}(x_1)$ with[8]

1. $B_{2\delta}(x_0), B_{2\delta}(x_1) \subset V$;
2. $B_{2\delta}(x_0) \cap B_{2\delta}(x_1) = \emptyset$;
3. $\delta < .$

Then the sets $B_\delta(x_0) = V_0$ and $B_\delta(x_1) = V_1$ are as in the claim.   CLAIM ⊣

CLAIM. There is a collection of open sets indexed by *finite* binary sequences, $(U_s)_{s \in \{0,1\}^{<\mathbb{N}}}$, such that

1. $U_s \cap P \neq \emptyset$ all $s \in \{0,1\}^{<\mathbb{N}}$;
2. $\overline{U_{s^\frown 0}}, \overline{U_{s^\frown 1}} \subset U_s$ all $s \in \{0,1\}^{<\mathbb{N}}$;[9]
3. $d(U_s) < \frac{1}{n}$ all $s \in \{0,1\}^n$;
4. $U_{s^\frown 0} \cap U_{s^\frown 1} = \emptyset$ all $s \in \{0,1\}^{<\mathbb{N}}$.

PROOF OF CLAIM. We build $(U_s)_{s \in \{0,1\}^{\leq n}}$ by induction on $n$. We can just begin with $U_\emptyset$ any sufficiently small open set meeting $P$.

Given the collection $(U_s)_{s \in \{0,1\}^n}$ we can choose for each $s$

$$U_{s^\frown 0}, U_{s^\frown 1} \subset U_s$$

as in the last claim for $= \frac{1}{n+1}$.   CLAIM ⊣

---

[6]$\overline{A}$ indicates the closure of a set $A$.

[7]Here and elsewhere I use $d(V)$ to denote the *diameter* of $V$, $\sup\{d(x,y): x, y \in V\}$, where $d$ is some complete metric on our Polish space.

[8]In general, $B_r(x)$ denotes the open ball of radius $r$ around $x$: $\{y : d(x,y) < r\}$.

[9]For $s : \{0, 1, \ldots, n-1\} \to \{0,1\}$ we use $s^\frown i$ to indicate the sequence of length $n+1$ obtained by concatenating $s$ with the sequence $\langle i \rangle$; that is to say, $s^\frown i$ extends $s$, has length equal to length $(s) + 1$, and assumes $i$ as its final value.

In this way we have obtained a nested collection of open sets whose structure resembles the canonical basis of open sets in the space $\{0, 1\}^{\mathbb{N}}$ of infinite binary sequences. For each $f : \mathbb{N} \to \{0, 1\}$ and $n \in \mathbb{N}$ we let $f|_n$ be the binary sequence of length $n$ obtained by restricting $f$ to its first $n$ many values.

CLAIM. For each $f : \mathbb{N} \to \{0, 1\}$ there is $x_f \in P$ with

$$x_f \in \bigcap_{n \in \mathbb{N}} U_{f|_n}.$$

PROOF OF CLAIM. We may choose $z_n \in U_{f|_n} \cap P$ for each $n$ by assumption on our nested collection of open sets. Then we obtain that $d(z_n, z_m) < \frac{1}{n}$ all $m \geq n$ since both points are in $U_{f|_n}$ and this open set has diameter less than $\frac{1}{n}$. Thus $(z_n)$ must be a Cauchy sequence, which will necessarily converge to some point

$$z_\infty \in P \cap \bigcap_{n \in \mathbb{N}} \overline{U_{f|_n}}.$$

Since $\overline{U_{f|_{n+1}}} \subset U_{f|_n}$ we obtain $x_f = z_\infty$ is as required.              CLAIM ⊣

For each such $f$ there will be exactly one $x_f$ as in the claim, for any other

$$x'_f \in \bigcap_{n \in \mathbb{N}} U_{f|_n}$$

would have to have $d(x_f, x'_f) < 1/n$ for all $n$.

Thus $f \mapsto x_f$ provides some kind of map from the space of infinite binary sequences to $P$. It is obviously one-to-one, since if $f \neq f'$ then we may let $s = f \cap f'$ be the longest initial segment contained in both, and obtain that one of $x_f, x_{f'}$ is in $U_{s \frown 0}$ and the other is in $U_{s \frown 1}$. The main issue is continuity.

CLAIM. $f \mapsto x_f$ is continuous.

PROOF OF CLAIM. Let $V$ be an open neighborhood of $x_f$. (We need an open $W$ containing $f$ with $\{x_h : h \in W\} \subset V$.) Choose $n$ with $B_{1/n}(x_f) \subset V$. Then since any two points $y_0, y_1$ of $U_{f|_n}$ have $d(y_0, y_1) < 1/n$ we obtain $U_{f|_n} \subset V$, and thus for $W$ the open set $\{h : \mathbb{N} \to \{0, 1\} \mid h \supset f|_n\}$ we have $x_h \in V$ all $h \in W$.              CLAIM ⊣

Thus

$$\{0, 1\}^{\mathbb{N}} \to P$$
$$f \mapsto x_f$$

is a continuous function. Since $\{0, 1\}^{\mathbb{N}}$ is compact the map sends closed sets to closed sets. Since it is injective we then have that it sends open sets to sets which are relatively open in its image, and thus it is a homeomorphism onto a closed subset of $P$.

$P$ is at least as big as the number of subsets of $\mathbb{N}$; thus $|P| \geq 2^{\aleph_0}$. But conversely, $P$ is a Polish space in its own right, and so it has a countable dense subset, $\mathcal{D}$ say; and then $P$ is no bigger than the number of Cauchy sequences from $\mathcal{D}$, which gives us the bound

$$|P| \leq |\mathcal{D}|^{\mathbb{N}} \leq |\mathcal{P}(\mathcal{D} \times \mathbb{N})| = |2^{\mathcal{D} \times \mathbb{N}}| = 2^{\aleph_0}. \qquad \dashv$$

COROLLARY 2.16. *Any uncountable Borel set has size* $2^{\aleph_0}$.

PROOF. We have previously seen that any Borel set allows a Polish topology and any uncountable Polish space contains a perfect subset and hence has cardinality continuum. $\qquad \dashv$

COROLLARY 2.17. *Any Polish space without isolated points has size* $2^{\aleph_0}$.

PROOF. As remarked above, a Polish space without isolated points must contain a perfect set. $\qquad \dashv$

Thus in particular $\mathbb{Q}$ in the subspace topology it inherits from $\mathbb{R}$ is not a Polish space.

We are yet to see that any two Borel sets of the same cardinality are Borel isomorphic. In light of our lemmas on changing topologies we need only see that any two Polish spaces of the same cardinality are isomorphic as standard Borel spaces. The proof of this is sketched in the exercises below.

EXERCISES 2.18.

1. Show that $\mathbb{N}^{\mathbb{N}}$ is homeomorphic to a $G_\delta$ subset of $\{0, 1\}^{\mathbb{N}}$. (Namely, the set of binary sequences with infinitely many 1s).

2. (a) Show that if $X$ is Polish then we may find a nested collection of sets $(V_s)_{s \in \mathbb{N}^{<\mathbb{N}}}$ such that

      (i) each $V_s$ is nonempty and open,
      (ii) $\overline{V_s} \subset V_t$ whenever $s$ strictly extends $t$,
      (iii) $V_\emptyset = X$,
      (iv) $d(V_s) < \frac{1}{n}$ whenever $s : \{0, 1, \ldots, n-1\} \to \mathbb{N}$, $n > 0$,
      (v) each $V_s$ equals $\bigcup_{k \in \mathbb{N}} V_{s^\frown k}$.

  (b) Conclude that any Polish space is the continuous, open image of $\mathbb{N}^{\mathbb{N}}$.

  (c) Show that if $X$ is Polish then there is a Borel $B \subset \mathbb{N}^{\mathbb{N}}$ which is Borel isomorphic to $X$.

3. Let $X_1, X_2$ be Polish spaces, $B_1 \subset X_1$, $B_2 \subset X_2$ Borel sets, $\pi_1 : X_1 \to B_2$, $\pi_2 : X_2 \to B_1$ Borel isomorphisms.

Show that $X_1$ and $X_2$ are Borel isomorphic. (Hint: This is close to the proof of Schroeder-Bernstein which does *not* use choice. You can find *that* proof in Tom Jech's *Set Theory*.)

4. Conclude from 1-3 above and the earlier results about perfect sets that any two uncountable Borel sets are Borel isomorphic.

5. Show that if $X$ is a countable Polish space then any subset is Borel. Thus any function between countable Polish spaces is Borel.

## §3. Borel equivalence relations.

DEFINITION 3.1. An equivalence relation $E$ on a Polish space $X$ is *Borel* if it is Borel as a subset of $X \times X$ in the product topological structure. A function

$$\theta : X \to Y$$

between Polish spaces is *Borel* if $\theta^{-1}[U]$ is Borel for any open set $U \subset Y$. (Actually this is equivalent to the graph of $\theta$ being Borel as a subset of $X \times Y$, but we are unlikely to need that particular result.)

For $E, F$ equivalence relations on $X, Y$ we write

$$E \leq_B F,$$

$E$ *Borel reducible to* $F$, if there is a Borel $\theta : X \to Y$ such that for all $x_1, x_2 \in X$

$$x_1 E x_2 \Leftrightarrow \theta(x_1) F \theta(x_2).$$

We can then go ahead and set $E <_B F$ if there is a Borel reduction of $E$ to $F$ but not the other way around, and $E \sim_B F$ if there is a Borel reduction in both directions.

Of course, there are other notions of comparison we might use for Borel equivalence relations. We might ask that the reduction be one-to-one. Or that it provide a bijection with some invariant Borel subset of the target space. And so on.

Although certainly not the only notion worth investigating, $\leq_B$ is the main one studied by descriptive set theorists. It, and its variant which allows classes of functions more general than Borel, also provides the most generous notion of reduction, and thus the nonreducibility below have the most force if we phrase them for $\leq_B$.

NOTATION 3.2. For $X$ a Polish space we let $\mathrm{id}(X)$ be the identity relation on $X$. (Some people use $\Delta(X)$ for this relation, which is suggested by its being the diagonal in $X \times X$.)

$E_0$ is the equivalence relation of agreement mod finite on $\{0, 1\}^{\mathbb{N}}$: $f_1 E_0 f_2$ if

$$\exists N \forall m > N \left( f_1(m) = f_2(m) \right).$$

From now on let me just write $2^{\mathbb{N}}$ for the space $\{0, 1\}^{\mathbb{N}}$. It is a little less cumbersome and formal.

LEMMA 3.3. $\mathrm{id}(2^{\mathbb{N}}) \leq_B E_0$.

PROOF. Let $(s_n)_{n \in \mathbb{N}}$ enumerate the set $\{0, 1\}^{<\mathbb{N}}$ of all finite binary sequences. For each $f \in 2^{\mathbb{N}}$ we define $\theta(f)$ by

$$(\theta(f))(n) = 1 \Leftrightarrow s_n \subset f.$$

That is, $(\theta(f))(n) = 1$ if $s_n$ is a substring of $f$.

This is continuous, since if $s_n : \{0, 1, \ldots, k-1\} \rightarrow \{0, 1\}$ and $(\theta(f_0))(n) = i$ ($i \in \{0, 1\}$), then for any other $f$ in the open set $\{h : \forall j < k(h(j) = f_0(j))\}$ we have $(\theta(f))(n) = i$.

And obviously if $f_1 \, \mathrm{id}(2^{\mathbb{N}}) f_2$ then by definition $f_1 = f_2$ and so $\theta(f_1) = \theta(f_2)$ and they are certainly $E_0$ equivalent.

Conversely if $f_1(k) \neq f_2(k)$ then for all $s_n$ with $s_n \subset f_1$ of length greater than $k$ we have $(\theta(f_1))(n) = 1$ while $(\theta(f_2))(n) = 0$.                        $\dashv$

To show failure of reducibility in the other direction I will use Baire category techniques. The idea here is to approximate Borel functions on some suitably "large" set by continuous functions.

DEFINITION 3.4. A subset of a space $X$ is *nowhere dense* if its intersection with any nonempty open subset of $X$ is not dense. A subset is *meager* if it is included in the countable union of sets which are each closed and nowhere dense. The complement of a meager set is said to be *comeager*.

FACT 3.5. The countable unions of meager sets are meager; countable intersections of comeager sets are comeager. The meager sets form a $\sigma$-ideal; the comeager sets form a $\sigma$-filter.

Some books only state the Baire category theorem for a few specific spaces like the unit interval. The usual proof extends to Polish spaces.

THEOREM 3.1 (Baire category theorem). *Any comeager subset of a Polish space is nonempty.*

COROLLARY 3.6. *If $X$ is a Polish space without isolated points then any comeager set is uncountable.*

DEFINITION 3.7. A set $A$ has *the Baire property* if for some open $\mathcal{O}$ the symmetric difference

$$A \Delta \mathcal{O} = (A \setminus \mathcal{O}) \cup (\mathcal{O} \setminus A)$$

is meager.

LEMMA 3.8. *Every Borel set has the Baire property.*

PROOF. Clearly the open sets have Baire property. If $C$ is closed then for

$$C^o = \bigcup \{U : U \subset C, U \text{ open}\}$$

we have that $C^o$ is the largest open set included in $C$ and $C \Delta C^o = C \setminus C^o$ is nowhere dense. Thus in general the complement of a set with Baire property has Baire property, since if $B \Delta \mathcal{O}$ is included in a meager set $M$, and $C$ is the complement of the open set $\mathcal{O}$, then

$$(X \setminus B) \Delta C^o \subset M \cup (C \setminus C^o).$$

Finally if $(B_i)_i$ is a sequence of Borel sets, $(\mathcal{O}_i)_i$ a sequence of open sets with $B_i \Delta \mathcal{O}_i$ always meager, then

$$\left(\bigcup_{i \in \mathbb{N}} B_i\right) \Delta \left(\bigcup_{i \in \mathbb{N}} \mathcal{O}_i\right)$$

is included in the countable union of meager sets, and is hence meager in its own right.                                                                       ⊣

COROLLARY 3.9. *Any Borel function between Polish spaces is continuous on a comeager set.*

That is, if $\theta : X \to Y$ is Borel, then there is a comeager $C \subset X$ having $\theta|_C$ continuous.

PROOF. Let $\{U_n \mid n \in \mathbb{N}\}$ be a countable basis of $Y$. For each $n$ choose open $\mathcal{O}_n \subset X$ with

$$\left(\theta^{-1}[U_n]\right) \Delta \mathcal{O}_n \subset M_n$$

for some meager set $M_n$. Let

$$C = X \setminus \left(\bigcup_{n \in \mathbb{N}} M_n\right).$$

Then we have that $\theta|_C$ pulls basic open sets back to relatively open subsets of $C$, and thus is continuous.                                               ⊣

LEMMA 3.10. *$E_0$ is not Borel reducible to $\mathrm{id}(2^\mathbb{N})$.*

PROOF. Instead let $\theta : 2^\mathbb{N} \to 2^\mathbb{N}$ be a Borel reduction. Let $C$ be a comeager set on which $\theta$ is continuous.

For each finite binary sequence $s \in \{0,1\}^{<\mathbb{N}}$ define

$$\psi_s : 2^\mathbb{N} \to 2^\mathbb{N}$$

by

$$(\psi_s(f))(n) = s(n) + f(n) \mod 2$$

if $n$ is less than the length of $s$, and equal to $f(n)$ otherwise.[10] Each $\psi_s$ is a homeomorphism, and thus $\psi_s(C)$ is again comeager.

Thus we can let

$$\hat{C} = \bigcap_{s \in 2^{<\mathbb{N}}} \psi_s[C]$$

---

[10] A way to think of this: Identify $\{0,1\}^\mathbb{N}$ with the product group $\mathbb{Z}_2^\mathbb{N}$ and consider the subgroup $G$ consisting of elements which have finite support. To each finite binary sequence $s$ we can associate a corresponding $g_s \in G$ and let $\psi_s$ be translation by $g_s$.

and obtain a comeager set which is invariant under $E_0$, in the sense that it includes any equivalence class it meets, and on which $\theta$ is continuous. Now choose $f \in \hat{C}$. The $E_0$-equivalence class

$$[f]_{E_0}$$

of $f$ is dense in $2^{\mathbb{N}}$ and included in $\hat{C}$; hence it is dense in $\hat{C}$. $\theta$ is constant on $[f]_{E_0}$ by the assumption it performs a reduction.

$\theta|_{\hat{C}}$ is continuous and constant on a dense subset, and therefore it is constant on $\hat{C}$. But this really does give us a contradiction, since $\hat{C}$ must be uncountable, and we can choose some

$$h \in \hat{C} \setminus [f]_{E_0}$$

with $\theta(h) = \theta(f)$. ⊣

EXERCISES 3.11.

1. Show that for any two uncountable Polish spaces $X$ and $Y$ we have $\mathrm{id}(X) \leq_B \mathrm{id}(Y)$.

2. Let $\mathcal{L}$ be a language consisting of just countably many unary predicates and let $F_2$ be the isomorphism relation on $\mathrm{Mod}(\mathcal{L})$. Show that $E_0 \leq_B F_2$.

Hence that $\mathrm{id}(\mathbb{R}) <_B F_2$.

The nonreducibility of $F_2$ to the identity relation on a Polish space even works for notions of reducibility which allow broader classes of functions. Consistently with ZFC $F_2$ is nonreducible to $\mathrm{id}(\mathbb{R})$ using "set recursive" or even "$\mathbb{R}$-ordinal definable" functions.[11] So although this isomorphism relation for countably many unary predicates might seem to be almost trivial from the point of view of model theory, there is already a substantial restriction on the kinds of objects one can hope to assign as complete invariants.

If you want a considerably harder exercise you might try showing that $E_0 <_B F_2$.

## §4. Infinitary logic.

DEFINITION 4.1. Let $\mathcal{L}$ be a language. $\mathcal{L}_{\omega_1,\omega}$ is obtained by closing the atomic formulas of $\mathcal{L}$ under the usual first-order operations

(a) $\psi \mapsto \exists x \psi$, $\psi \mapsto \forall x \psi$,
(b) $(\psi, \phi) \mapsto \psi \vee \phi$, $(\psi, \phi) \mapsto \psi \wedge \phi$,
(c) $\psi \mapsto \neg \psi$,

---

[11] I only say "consistently with ZFC" since we need to avoid, for instance, certain things which can happen in Gödel's $L$ as a result of the simply definable well order of $\mathbb{R}$. Certainly with a little bit of forcing one can guarantee all $\underset{\sim}{\Delta}^1_2$ sets have Baire property and with quite a lot of forcing one obtains the same for the $\mathbb{R}$-ordinal definable sets.

as well as infinitary conjunction and disjunction,

(d)  whenever $\{\psi_i \mid i \in \mathbb{N}\}$ is a countable set of formulas in $\mathcal{L}_{\omega_1,\omega}$ we have

$$\bigvee_{i\in\mathbb{N}} \psi_i \in \mathcal{L}_{\omega_1,\omega},$$

$$\bigwedge_{i\in\mathbb{N}} \psi_i \in \mathcal{L}_{\omega_1,\omega}.$$

DEFINITION 4.2.  If $\mathcal{M}$ is an $\mathcal{L}$-structure, and $\varphi(x_1, x_2, \ldots, x_k)$ is a formula of $\mathcal{L}_{\omega_1,\omega}$ with just finitely many free variables, then we can define

$$\mathcal{M} \models \varphi(a_1, \ldots, a_k)$$

by induction on the complexity of $\varphi$ in the usual way. The only clauses which differ from first-order logic are the ones corresponding to infinite conjunction and disjunction, and here we stipulate that

$$\mathcal{M} \models \bigwedge_{i\in\mathbb{N}} \psi_i \Leftrightarrow (\forall i \in \mathbb{N} (\mathcal{M} \models \psi_i)),$$

and

$$\mathcal{M} \models \bigvee_{i\in\mathbb{N}} \psi_i \Leftrightarrow (\exists i \in \mathbb{N} (\mathcal{M} \models \psi_i)).$$

TECHNICAL REMARKS.

(i)  The infinite conjunction $\bigvee_{i\in\mathbb{N}} \psi_i \in \mathcal{L}_{\omega,\omega}$ and infinite disjunction $\bigwedge_{i\in\mathbb{N}} \psi_i \in \mathcal{L}_{\omega,\omega}$ depend solely on the *set* $\{\psi_i \mid i \in \mathbb{N}\}$, and not the particular enumeration $(\psi_i)_{i\in\mathbb{N}}$.

(ii)  I have allowed formulas with infinitely many free variables, but not given any role to them in the satisfaction relation. There seems to be some variation in the literature on this point.

(iii)  Later we will need to perform a range of computations with languages and formulas, and for these purposes it will be helpful to adopt some conventions. First that we assume $\mathcal{L}$ consists solely of symbols which are themselves hereditarily countable[12] sets; all our languages will be countable, so this is not overly restrictive. After that we can assume that $\mathcal{L}_{\omega_1,\omega}$ is a subset of the hereditarily countable sets; for instance, we can have as our convention that $\bigvee$ is some fixed hereditarily countable set, such as the number 17, and that

$$\bigvee_{i\in\mathbb{N}} \psi_i = \left\{ \left\{ \psi_i, \bigvee \right\} : i \in \mathbb{N} \right\}.$$

(iv)  I will only use countably many variable symbols. Keisler handles this point somewhat differently.

---

[12]A set $A$ is *hereditarily countable* if it is countable, all its elements are countable, all their elements are countable, and so on. More precisely, HC, the *collection of hereditarily countable sets*, is the smallest set containing all its countable subsets.

DEFINITION 4.3. $F \subset \mathcal{L}_{\omega_1,\omega}$ is a *fragment* if
(a) it contains the atomic formulas;
(b) it is closed under the first-order operations

$$\psi \mapsto \exists x \psi,$$

$$\psi \mapsto \forall x \psi,$$

$$(\psi, \phi) \mapsto \psi \vee \phi,$$

$$(\psi, \phi) \mapsto \psi \wedge \phi,$$

$$\psi \mapsto \neg \psi;$$

(c) it is closed under subformulas, in the sense that

$$\exists x \psi \in F \Rightarrow \psi \in F,$$

$$\forall x \psi \in F \Rightarrow \psi \in F,$$

$$\psi \wedge \phi \in F \Rightarrow \psi, \phi \in F,$$

$$\psi \vee \phi \in F \Rightarrow \psi, \phi \in F,$$

$$\neg \psi \in F \Rightarrow \psi \in F,$$

$$\bigwedge_{i \in \mathbb{N}} \psi \in F \Rightarrow \{\psi_i : i \in \mathbb{N}\} \subset F,$$

$$\bigvee_{i \in \mathbb{N}} \psi \in F \Rightarrow \{\psi_i : i \in \mathbb{N}\} \subset F;$$

(d) if $\varphi(x_1, x_2, x_3, \dots) \in F$ and $t$ is a term, then

$$\varphi(t, x_2, x_3, \dots) \in F.$$

In this definition of fragment (d) is important for what it *does not* do. It does not license substitutions which replace infinitely many variables with terms. Otherwise all our fragments would be uncountable.

In Ramez Sami's "Polish group actions and the Vaught conjecture" he uses a somewhat weaker notion of fragment. As it turns out this weaker notion is still sufficient to prove that countable fragments generate Polish topologies, but not quite sufficient for us to obtain a nice characterization of *atomic model relative to a given fragment*.

DEFINITION 4.4. For $F \subset \mathcal{L}_{\omega_1,\omega}$ a fragment we let $\tau_F$ be the topology on $\mathrm{Mod}(\mathcal{L})$ whose basis consists of open sets of the form

$$\{\mathcal{M} \in \mathrm{Mod}(\mathcal{L}): \mathcal{M} \models \varphi(n_1, n_2, \dots, n_k)\},$$

where $n_1, n_2, \dots, n_k \in \mathbb{N}$ and $\varphi(\vec{x}) \in F$.

LEMMA 4.5. *If $F$ is a countable fragment then* $(\mathrm{Mod}(\mathcal{L}), \tau_F)$ *is a Polish space.*

PROOF. This follows our proof for the topology generated by first-order logic (see proof of Lemma 2.10). We let $\hat{\mathcal{L}}$ be the language obtained by adding fresh constants $(c_n)_{n \in \mathbb{N}}$ to $\mathcal{L}$; and then we obtain $\hat{F}$ as the fragment generated by $F$ in $\hat{\mathcal{L}}_{\omega_1, \omega}$. And parallel to the proof of Lemma 2.10 we let $\hat{S}$ be the set of sentences in $\hat{F}$.

$$2^{\hat{S}} =_{df} \{0, 1\}^{\hat{S}}$$

gives us a Polish space, and we can let $A_{\hat{S}}$ be the set of $f \in 2^{\hat{S}}$ satisfying

(a) any finite subset of $\{\psi : f(\psi) = 1\}$ is *consistent*, in the sense that it has some model;

(b) $f(\psi) = 1$ if and only if $f(\neg\psi) = 0$;

(c) $f(\exists x \psi(x)) = 1$ if and only if there is some $n$ with $f(\psi(c_n)) = 1$;

(d) $f(\bigvee_{i \in \mathbb{N}} \psi_i) = 1$ if and only if there is some $i \in \mathbb{N}$ with $f(\psi_i) = 1$.

We had to modify (a) a bit so that it still corresponds to a closed condition. We also had to add a condition (d), which is $G_\delta$ by the same calculation we previously used for (c). With these modifications one can show that if we assign to each $\mathcal{M} \in \mathrm{Mod}(\mathcal{L})$ the characteristic function of $F$-theory of its canonical expansion to a model of $\hat{\mathcal{L}}$, $T_{\mathcal{M}}$ where

$$T_{\mathcal{M}}\left(\psi\left(c_{n_1}, c_{n_2}, \ldots, c_{n_k}\right)\right) = 1 \Leftrightarrow \mathcal{M} \models \psi\left(n_1, n_2, \ldots, n_k\right),$$

then

$$\left(\mathrm{Mod}(\mathcal{L}), \tau_F\right) \to A_{\hat{F}}$$
$$\mathcal{M} \mapsto T_{\mathcal{M}}$$

provides a homeomorphism.                                                     ⊣

DEFINITION 4.6. If $\sigma \in \mathcal{L}_{\omega_1, \omega}$ is a sentence then we let $\mathrm{Mod}(\sigma)$ be the collection of $\mathcal{M} \in \mathrm{Mod}(\mathcal{L})$ satisfying $\sigma$; we let $F(\sigma)$ be the fragment generated by $\sigma$, and obtain from the last lemma that $\mathrm{Mod}(\sigma)$ is a Polish space in the subspace topology bequeathed by $\tau_{F(\sigma)}$.

If $F$ is a countable fragment and $T \subset F$ a theory, then we let $\mathrm{Mod}(T)$ be the models of $T$. This is a closed subset of $(\mathrm{Mod}(\mathcal{L}), \tau_F)$, and hence a Polish space in the subspace topology.

DEFINITION 4.7. Given a countable fragment $F$ we can define an *n-type over $F$* to be some subset of the $F$-formulas in free variables $x_1, x_2, \ldots, x_n$, and more generally we can define an *$F$-type* to be an $n$-type over $F$ for some $n$. A type is *$F$-complete* if any $\psi(\vec{x}) \in F$ in the appropriate variables is either in the type or has its negation in $F$. We then say that a complete type is *principal* over a complete theory $T \subset F$ if it contains some $\varphi(\vec{x})$ such that for all other and for all $\psi(\vec{x}) \in F$ we have either

$$\forall \vec{x}\left(\varphi(\vec{x}) \to \psi(\vec{x})\right)$$

or

$$\forall \vec{x}\left(\varphi(\vec{x}) \to \neg\psi(\vec{x})\right).$$

Given a model $\mathcal{M}$ and $\vec{a} \in \mathcal{M}$ we let the *F-type over* $\mathcal{M}$ be the set of $\psi(\vec{x})$: $\mathcal{M} \models \psi(\vec{a})$. $\mathcal{M}$ *omits* an $F$-type if there is no $\vec{a}$ which has that type as its $F$-type over $\mathcal{M}$. For $F$ a fragment $\mathrm{Th}_F(\mathcal{M})$ is the set of sentences in $F$ satisfied by $\mathcal{M}$. We say that a model $\mathcal{M}$ is *$F$-atomic* if it realizes no nonprincipal types over $\mathrm{Th}_F(\mathcal{M})$.

With these definitions out of the way, many of the usual facts for first-order logic generalize easily.

LEMMA 4.8. *Let $F$ be a countable fragment, $T \subset F$ a complete theory. Any two countable $F$-atomic models of $T$ are isomorphic.*

LEMMA 4.9. *Let $F$ be a countable fragment. Let $\Sigma(x_1, x_2, \ldots, x_n)$ be a complete nonprincipal type over $F$ and let $k_1, k_2, \ldots, k_n \in \mathbb{N}$. Let $T \subset F$ be a complete theory. Then the set of $\mathcal{M} \in \mathrm{Mod}(T)$ with*

$$\mathcal{M} \models \Sigma(k_1, k_2, \ldots, k_n)$$

*is closed nowhere dense (with respect to $\tau_F$).*

COROLLARY 4.10 (Omitting types theorem). *$F, T, \Sigma(\vec{x})$ as above. The set of $\mathcal{M}$ in $(\mathrm{Mod}(T), \tau_F)$ omitting $\Sigma(\vec{x})$ is comeager (in $(\mathrm{Mod}(T), \tau_F)$).*

LEMMA 4.11. *Let $F$ be a countable fragment, $T \subset F$ a complete theory, $\mathcal{M}_0 \in \mathrm{Mod}(T)$. The set of $\mathcal{N} \in (\mathrm{Mod}(T), \tau_F)$ isomorphic to $\mathcal{M}_0$ is comeager if and only if $\mathcal{M}_0$ is atomic.*

PROOF. First suppose that $\mathcal{M}_0$ is atomic.

CLAIM. For each $\vec{n} \in \mathbb{N}^{<\mathbb{N}}$ the set of $\mathcal{N} \in (\mathrm{Mod}(T), \tau_F)$ with $\vec{n}$ realizing a principal type is open dense.

PROOF OF CLAIM. The set is immediately open by the definition of the topology. To see density, let us consider some nonempty basic open set of the form

$$V = \{\mathcal{N} : \mathcal{N} \models \psi(\vec{n}, \vec{m})\}$$

where $\vec{m} \in \mathbb{N}^{<\mathbb{N}}$ has been chosen to be disjoint from $\vec{n}$ and $\psi(\vec{x}, \vec{y}) \in F$. Then we can consider the formula

$$\exists \vec{x} \theta(\vec{x}) =_{\mathrm{df}} \exists \vec{x} \exists \vec{y} \psi(\vec{x}, \vec{y}) \wedge \bigwedge_{i < lh(\vec{x})} \bigwedge_{j < lh(\vec{y})} x_i \neq y_j.$$

Since this is realized by some tuple in some model in our space, and since $\mathcal{M}_0$ satisfies the complete theory $T$, we may find some $\vec{a} \in \mathcal{M}_0$ with

$$\mathcal{M}_0 \models \exists \vec{y} \psi(\vec{a}, \vec{y}) \wedge \bigwedge_{i < lh(\vec{x})} \bigwedge_{j < lh(\vec{y})} a_i \neq y_j.$$

We then find some $\varphi(\vec{x})$ with

$$\mathcal{M}_0 \models \varphi(\vec{a})$$

witnessing that $\vec{a}$ realizes some principal type. But then the set

$$U = \left\{ \mathcal{N} : \mathcal{N} \models \psi(\vec{n}, \vec{m}) \wedge \varphi(\vec{n}) \wedge \bigwedge_{i < lh(\vec{x})} \bigwedge_{j < lh(\vec{y})} n_i \neq m_j \right\}$$

is an open subset of $V$ of whose elements have $\vec{n}$ realizing a principal type; to see that this set is nonempty we choose some $\vec{b} \in \mathcal{M}_0$ disjoint from $\vec{a}$ with $\mathcal{M}_0 \models \varphi(\vec{a}, \vec{b})$ and some isomorphic copy of $\mathcal{M}_0$ where $\vec{n}$ goes to $\vec{a}$ and $\vec{m}$ goes to $\vec{b}$.      CLAIM ⊣

Thus we have that the set of atomic models is comeager, and since any two countable atomic models are isomorphic we are done.

Conversely, if $\mathcal{M}_0$ is *not* atomic, then we can consider some principal complete type $\Sigma(\vec{x})$ realized by $\mathcal{M}_0$. By the omitting types theorem the collection of models omitting $\Sigma(\vec{x})$ is comeager. By the Baire category theorem we have that this set intersects any comeager set.      ⊣

LEMMA 4.12. *Let $F$ be a countable fragment, $T \subset F$ a complete theory, $\mathcal{M}_0 \in \mathrm{Mod}(T)$. If there is no $F$-atomic model of $T$ then the isomorphism relation is meager as a subset of*

$$\left( \mathrm{Mod}(T), \tau_F \right) \times \left( \mathrm{Mod}(T), \tau_F \right).$$

PROOF. It is easily seen that the nonexistence of an atomic model is equivalent to the principal types not being dense in the space of $n$-types[13] for some $n$. So let us choose some $\psi(\vec{x})$ which extends to no principal type. Then we have that for any $\vec{n}, \vec{m}$ the collection $U_{\vec{m}\vec{n}}$ of $(\mathcal{M}, \mathcal{N})$ with

$$\mathcal{M} \models \psi(\vec{m}),$$

$$\mathcal{N} \models \psi(\vec{n}),$$

is relatively clopen in the product space; but the subset $C_{\vec{m}, \vec{n}}$ consisting of $(\mathcal{M}, \mathcal{N})$ over which $\vec{m}$ and $\vec{n}$, respectively, realize the same type is a closed nowhere dense subset of $U_{\vec{m}, \vec{n}}$.

But then we have for any $(\mathcal{M}, \mathcal{N})$ in the comeager set

$$\bigcap_{\vec{m}, \vec{n}} \left( \left( \mathrm{Mod}(T), \tau_F \right) \times \left( \mathrm{Mod}(T), \tau_F \right) \setminus C_{\vec{m}, \vec{n}} \right)$$

there must, by completeness of $T$, be *some* $\vec{k}, \vec{\ell}$ with

$$\mathcal{M} \models \psi(\vec{k}),$$

$$\mathcal{N} \models \psi(\vec{\ell}).$$      ⊣

---

[13]We define the space of $n$-types in the fragment $F$ to be the collection of all complete consistent $F$-types $\Sigma(x_1, x_2, \ldots, x_n)$ in some fixed string of $n$ variables $x_1, \ldots, x_n$, with the topology generated by the basic open sets of the form $\{\Sigma(\vec{x}) : \varphi(\vec{x}) \in \Sigma(\vec{x})\}$ for some $\varphi \in F$. It is easily seen that in this topology the isolated points are exactly the atomic types.

The above lemma is the main fact we will need about infinitary logic. There is one other result worth mentioning just for comparison. This further result requires another definition.

DEFINITION 4.13. $S_\infty$ is the group of all permutations of the natural numbers with topology of pointwise convergence. Thus a basic open set consists of all bijections $\pi\colon \mathbb{N} \to \mathbb{N}$ with

$$\forall i < k\left(\pi(i) = n_i\right),$$

for some $k, n_1, n_2, \ldots, n_k \in \mathbb{N}$.

$S_\infty$ is a topological group, since the group operations are all continuous; it is also $G_\delta$ as a subset of $\mathbb{N}^{\mathbb{N}}$, and hence Polish as a space.

Note that $S_\infty$ acts continuously and in a canonical fashion on any $(\mathrm{Mod}(\mathcal{L}), \mathcal{L})$; and moreover its orbit equivalence relation is the isomorphism relation.

THEOREM 4.1 (Becker-Kechris). *Let $X$ be a standard Borel space on which $S_\infty$ acts in a Borel manner.*[14] *Then there is some countable $\mathcal{L}$ and sentence $\sigma \in \mathcal{L}_{\omega_1,\omega}$ and Borel isomorphism*

$$\theta\colon X \to \mathrm{Mod}(\sigma)$$

*intertwining the actions—that is, for all $\pi \in S_\infty$ and $x \in X$*

$$\pi \cdot \theta(x) = \theta(\pi \cdot x).$$

EXERCISES 4.14.

1. Show that if $T$ is a complete theory for the countable fragment $F$ then every isomorphism type in $\mathrm{Mod}(T)$ is dense. That is, if $\mathcal{M} \in \mathrm{Mod}(T)$, then the set of $\mathcal{N} \in \mathrm{Mod}(T)$ with

$$\mathcal{M} \cong \mathcal{N}$$

is dense relative to $\tau_F$.

2. $T, F$ as above. Show that if the isomorphism type of $\mathcal{M}$ is $G_\delta$, then it is comeager; and moreover if $\mathcal{M}$ is atomic then its isomorphism type is $G_\delta$.

3. $T, F$ as above. Show that canonical action of $S_\infty$ on $(\mathrm{Mod}(T), \tau_F)$ is continuous.

4. $T, F$ as above. For $\mathcal{M} \in (\mathrm{Mod}(T), \tau_F)$ we have that the induced map

$$S_\infty \to \{\mathcal{N} \in \mathrm{Mod}(T)\colon \mathcal{N} \cong \mathcal{M}\}$$

$$\pi \mapsto \pi \cdot \mathcal{M}$$

is open if and only if $\mathcal{M}$ is $F$-atomic.

---

[14]I.e., the map

$$(\pi, x) \mapsto \pi \cdot x,$$
$$S_\infty \times X \to X$$

is Borel.

5. Conclude that for $T$, $F$ as above the following are equivalent for $\mathcal{M} \in (\text{Mod}(T), \tau_F)$:

(a) $\mathcal{M}$ is atomic;

(b) the isomorphism type of $\mathcal{M}$ is comeager;

(c) the isomorphism type of $\mathcal{M}$ is $G_\delta$;

(d) the map $\pi \mapsto \pi \cdot \mathcal{M}$ is open as a map to the isomorphism type of $\mathcal{M}$.

For an arbitrary continuous action of a Polish group $G$ on a Polish space Ed Effros has shown that an orbit $[x]$ is $G_\delta$ if and only if it is comeager in its closure, if and only if the natural map $g \mapsto g \cdot x$ is open.

## §5. Scott's analysis.

DEFINITION 5.1. For $\mathcal{M}$ a model and $\vec{a}$ a finite sequence from $\mathcal{M}$ we define

$$\varphi_\alpha^{\vec{a}, \mathcal{M}}$$

by induction on the ordinal $\alpha$.

$$\varphi_0^{\vec{a}, \mathcal{M}} = \bigwedge \{\psi(\vec{x}) : \psi(\vec{x}) \text{ quantifier free}, \ \mathcal{M} \models \psi(\vec{a})\}$$

$$\varphi_{\alpha+1}^{\vec{a}, \mathcal{M}} = \varphi_\alpha^{\vec{a}, \mathcal{M}} \bigwedge \left\{ \exists \vec{y} \varphi_\alpha^{\vec{a}\vec{b}, \mathcal{M}} (\vec{x}, \vec{y}) : \vec{b} \in \mathcal{M}^{<\mathbb{N}} \right\}$$

$$\wedge \bigwedge_{n \in \mathbb{N}} \forall y_0, y_1, \ldots, y_n \bigvee \left\{ \varphi_\alpha^{\vec{a}\vec{b}, \mathcal{M}} (\vec{x}, \vec{y}) : \vec{b} \in \mathcal{M}^{n+1} \right\}.$$

For $\lambda$ a limit we set

$$\varphi_\lambda^{\vec{a}, \mathcal{M}} = \bigwedge_{\alpha < \lambda} \varphi_\alpha^{\vec{a}, \mathcal{M}}.$$

Note then that if $\mathcal{M}$ is a countable model of a countable language $\mathcal{L}$ then we have for each $\alpha < \omega_1$ and $\vec{a} \in \mathcal{M}$

$$\varphi_\alpha^{\vec{a}, \mathcal{M}} \in \mathcal{L}_{\omega_1, \omega}.$$

For $\alpha < \beta$

$$\varphi_\beta^{\vec{a}, \mathcal{M}} = \varphi_\beta^{\vec{b}, \mathcal{M}} \Rightarrow \varphi_\alpha^{\vec{a}, \mathcal{M}} = \varphi_\alpha^{\vec{b}, \mathcal{M}}.$$

LEMMA 5.2. *For $\mathcal{M}$, $\mathcal{L}$ countable there will be an ordinal $\delta < \omega_1$ such that for all $\vec{a}, \vec{b} \in \mathcal{M}$*

$$\exists \gamma < \omega_1 \left( \varphi_\gamma^{\vec{a}, \mathcal{M}} \neq \varphi_\gamma^{\vec{b}, \mathcal{M}} \right) \Leftrightarrow \varphi_\delta^{\vec{a}, \mathcal{M}} \neq \varphi_\delta^{\vec{b}, \mathcal{M}}.$$

In general such an ordinal $\delta$ with the property that

$$\forall \vec{a}, \vec{b} \in \mathcal{M} \left( \exists \gamma \left( \varphi_\gamma^{\vec{a}, \mathcal{M}} \neq \varphi_\gamma^{\vec{b}, \mathcal{M}} \right) \Leftrightarrow \left( \varphi_\delta^{\vec{a}, \mathcal{M}} \neq \varphi_\delta^{\vec{b}, \mathcal{M}} \right) \right)$$

will be guaranteed to exist by replacement, as we may take the supremum over $\vec{a}, \vec{b} \in \mathcal{M}$ of the least $\alpha$ (when it exists) with

$$\varphi_\alpha^{\vec{a}, \mathcal{M}} \neq \varphi_\alpha^{\vec{b}, \mathcal{M}}.$$

However it may be uncountable. For instance taking the usual ordering on ordinals, the Scott height of the structure

$$(\omega_1, <)$$

is itself $\omega_1$.

DEFINITION 5.3. For $\mathcal{M}$ a countable model the least $\delta$ as in Lemma 5.2 is the *Scott height of* $\mathcal{M}$ and will be denoted by $\alpha(\mathcal{M})$.

$$\varphi^{\emptyset,\mathcal{M}}_{\alpha(\mathcal{M})+2} =_{df} \varphi^{\mathcal{M}}$$

is the (canonical) *Scott sentence of* $\mathcal{M}$;

$$\varphi^{\vec{a},\mathcal{M}}_{\alpha(\mathcal{M})+2} =_{df} \varphi^{\vec{a},\mathcal{M}}$$

(or just $\varphi^{\vec{a}}$ when context indicates $\mathcal{M}$) is the *Scott sentence of* $\vec{a}$ (in $\mathcal{M}$); each $\varphi^{\vec{a},\mathcal{M}}_\alpha$ is the $\alpha$*th approximation of the Scott sentence for* $\vec{a}$ or *the canonical* $\alpha$-*type of* $\vec{a}$.

Note that if $\vec{a}, \vec{b} \in \mathcal{M}$ with

$$\varphi^{\vec{a},\mathcal{M}}_{\alpha(\mathcal{M})} = \varphi^{\vec{b},\mathcal{M}}_{\alpha(\mathcal{M})},$$

then it follows from the definition of $\alpha(\mathcal{M})$ that, in particular, we must have

$$\varphi^{\vec{a},\mathcal{M}}_{\alpha(\mathcal{M})+1} = \varphi^{\vec{b},\mathcal{M}}_{\alpha(\mathcal{M})+1}.$$

LEMMA 5.4. *If* $\varphi^{\vec{a},\mathcal{M}}_\beta \neq \varphi^{\vec{b},\mathcal{M}}_\beta$ *then in any other model* $\mathcal{N}$

$$\mathcal{N} \models \forall \vec{x} \left( \varphi^{\vec{a},\mathcal{M}}_\beta(\vec{x}) \Rightarrow \neg \varphi^{\vec{b},\mathcal{M}}_\beta (\vec{x}) \right).$$

PROOF. We prove this by induction on $\beta$. It is obvious for $\beta = 0$, and the passage through limits follows trivially from the structure of the definition.

For the inductive step, assume the lemma is true at $\beta$ and without loss that

$$\mathcal{N} \models \varphi^{\vec{a},\mathcal{M}}_{\beta+1} (\vec{c})$$

and that there is some $\vec{d} \in \mathcal{M}$ with

$$\forall \vec{e} \in \mathcal{M} \left( \varphi^{\vec{a}\vec{d},\mathcal{M}}_\beta \neq \varphi^{\vec{b}\vec{e},\mathcal{M}}_\beta \right).$$

Then we may find some $\vec{f} \in \mathcal{N}$ with

$$\mathcal{N} \models \varphi^{\vec{a}\vec{d}}_\beta(\vec{c}, \vec{f}),$$

and hence for all $\vec{e} \in \mathcal{M}$ we have

$$\mathcal{N} \models \neg \varphi^{\vec{b}\vec{e},\mathcal{M}}_\beta(\vec{c}, \vec{f}). \qquad \dashv$$

THEOREM 5.1. (*Scott*) *Any two countable models satisfying the same Scott sentence are isomorphic.*

PROOF. Let $\mathcal{M}, \mathcal{N}$ be countable models with

$$\mathcal{N} \models \varphi^{\mathcal{M}}.$$

Fix some ordinal $\alpha = \alpha(\mathcal{M})$ as the Scott height of $\mathcal{M}$.

CLAIM. For all $\vec{a} \in \mathcal{N}, \vec{b} \in \mathcal{M}$

$$\mathcal{N} \models \varphi_\alpha^{\vec{b},\mathcal{M}}(\vec{a})$$

implies that

$$\mathcal{N} \models \varphi_{\alpha+1}^{\vec{b},\mathcal{M}}(\vec{a}).$$

PROOF OF CLAIM. Otherwise from the assumption that $\mathcal{N}$ satisfies the Scott sentence of $\mathcal{M}$ we have

$$\mathcal{N} \models \forall \vec{x} \bigvee_{\vec{c} \in \mathcal{M}} \varphi_{\alpha+1}^{\vec{c},\mathcal{M}}(\vec{x}).$$

Hence we may find some $\vec{c} \in \mathcal{M}$ with

$$\mathcal{N} \models \varphi_{\alpha+1}^{\vec{c},\mathcal{M}}(\vec{a}),$$
$$\mathcal{N} \models \varphi_\alpha^{\vec{c},\mathcal{M}}(\vec{a})$$

and so by Lemma 5.2 we must have

$$\varphi_\alpha^{\vec{c},\mathcal{M}} = \varphi_\alpha^{\vec{b},\mathcal{M}},$$
$$\mathcal{M} \models \varphi_\alpha^{\vec{c},\mathcal{M}}(\vec{b})$$

but by the assumption the claim fails

$$\mathcal{M} \models \neg\varphi_{\alpha+1}^{\vec{c},\mathcal{M}}(\vec{b}),$$

with a contradiction to $\alpha$ being the Scott height of $\mathcal{M}$.          CLAIM ⊣

But now we can just go ahead with the usual back and forth kind of argument. We can define a sequence

$$a_0, a_1, a_2, \cdots \in \mathcal{M},$$
$$b_0, b_1, b_2, \cdots \in \mathcal{N},$$

with at each $k$

$$\mathcal{N} \models \varphi^{\langle a_0, a_1, \ldots, a_k \rangle, \mathcal{M}}(b_0, b_1, \ldots, b_k)$$

and under some fixed enumerations of the two models we have that the $i$th element of $\mathcal{M}$ appears among $\{a_0, a_1, \ldots, a_{2i}\}$ and the $i$th element of $\mathcal{N}$ appears among the $\{b_0, b_1, \ldots, b_{2i+1}\}$. We are able to propagate this construction since the claim above, in particular, implies that if $\mathcal{M} \models \varphi_\alpha^{\vec{b},\mathcal{N}}(\vec{a})$ and $d \in \mathcal{N}$ then

$$\mathcal{M} \models \exists y \varphi_\alpha^{\vec{b}^\frown d, \mathcal{N}}(\vec{a}, y),$$

which is the same as saying there is some $d' \in \mathcal{M}$ with

$$\mathcal{M} \models \varphi_\alpha^{\vec{b}^\frown d, \mathcal{N}} (\vec{a}, d'). \qquad\qquad \dashv$$

This was all proved because Dana Scott wanted to show the following.

COROLLARY 5.5. *The isomorphism type of any countable model is Borel in* Mod($\mathcal{L}$).

It has previously been shown that the isomomorphism type of any *countable ordinal* is Borel in Mod($\{<\}$); indeed it is a nice exercise to prove this by induction on the ordinal, without any appeal to the Scott analysis. With this result in mind Kuratowski had asked whether the isomorphism type of any countable structure was Borel, which in turn was explicitly answered as above.

EXERCISES 5.6.

1. $F$ a countable fragment and $T$ an $F$-complete theory. The isomorphism type of any $\mathcal{M} \in (\text{Mod}(T), \tau_F)$ has Baire property and is nonmeager if and only if it is comeager.

2. Let $\mathcal{L}$ be a countable language and $(F_\alpha)_{\alpha \in \omega_1}$ be an increasing sequence of countable fragments in $\mathcal{L}_{\omega_1, \omega}$. Suppose $\mathcal{M} \in \text{Mod}(\mathcal{L})$ is fortunate enough that for each $\alpha$ and $\vec{a} \in \mathcal{M}$ there is some $\psi(\vec{x}) \in F_{\alpha+1}$ such that

$$\mathcal{M} \models \psi(\vec{a})$$

$$\mathcal{M} \models \forall \vec{x} \forall \vec{y} \bigwedge_{\varphi \in F_\alpha} (\psi(\vec{x}) \wedge \psi(\vec{y}) \Rightarrow (\varphi(\vec{x}) \Leftrightarrow \varphi(\vec{y})));$$

in other words, for each $\vec{a}$ there is some formula in $F_{\alpha+1}$ satisfied by $\vec{a}$ which "isolates its type over $F_\alpha$".

Then show that there is some countable $\gamma < \omega_1$ with $\mathcal{M}$ atomic relative to the fragment $F_\gamma$ and the theory $\text{Th}_{F_\gamma}(\mathcal{M})$.

3. Show that if $\delta$ is the Scott height of a countable model then we have *for all* ordinals $\gamma$

$$\varphi_\gamma^{\vec{a}, \mathcal{M}} \neq \varphi_\gamma^{\vec{b}, \mathcal{M}} \Rightarrow \varphi_\delta^{\vec{a}, \mathcal{M}} \neq \varphi_\delta^{\vec{b}, \mathcal{M}}.$$

4. (a) Show that if $T \subset F$ is a complete theory then

$$\left\{ \varphi_0^{\vec{a}, \mathcal{M}} : \mathcal{M} \models T, \vec{a} \in \mathcal{M} \right\}$$

is a Borel set in some standard Borel structure; and hence has size continuum or $\leq \aleph_0$.

(b) Show that if $\sigma$ is a sentence appearing in some fragment $F$ then the collection of countable complete "consistent"[15] $T \subset F$ containing $\sigma$ is a Borel set.

---

[15]Of course, I have been trying to avoid use of this word in the context of infinitary model theory. Here I mean that the theory has some model.

(c) Conclude that for any $\sigma \in \mathcal{L}_{\omega_1,\omega}$

$$\left\{ \varphi_0^{\vec{a},\mathcal{M}} : \mathcal{M} \models \sigma, \vec{a} \in \mathcal{M} \right\}$$

is either countable or has cardinality $2^{\aleph_0}$.

(d) Iterating the argument of (a)–(c) show that for any $\sigma \in \mathcal{L}_{\omega_1,\omega}$ and $\alpha < \omega_1$

$$\left\{ \varphi_\alpha^{\vec{a},\mathcal{M}} : \mathcal{M} \models \sigma, \vec{a} \in \mathcal{M} \right\}$$

is either countable or has cardinality $2^{\aleph_0}$.

(e) (Morley's lemma) Show that if $\sigma$ has less than continuum many non-isomorphic countable models then it has $\leq \aleph_1$ many nonisomorphic countable models.

Actually the proof of the main theorem we are heading toward parallels the above argument due to Morley. There again is a kind of transfinite induction, where the failure at each level for their being "lots of models" gives us the ability to perform one further step of the analysis.

In Morley's lemma "lots of models" corresponds to $2^{\aleph_0}$ many countable models; and the technical tool used at each step is the perfect set theorem for Borel sets. In our setting the "lots of models" case will correspond to an embedding of $E_0$ and the central technical device is a version of Glimm-Effros due to Becker and Kechris.

## §6. Glimm-Effros.

THEOREM 6.1 (Becker-Kechris). *Let $E$ be an equivalence relation on a Polish space $X$ and let $G$ be a group acting by homeomorphisms on $X$. Then assuming*

(a) *$E$ is meager as a subset of $X \times X$, and*
(b) *for all $g \in G, x \in X$ we have*

$$g \cdot xEx$$

*(that is, the orbit equivalence relation induced by $G$ is included in $E$), and*
(c) *there is some $x_0 \in X$ with $[x_0]_G =_{df} \{g \cdot x_0 \mid g \in G\}$ dense,*

*then*

$$E_0 \leq_B E.$$

PROOF. Fix a complete metric $d$ on $X$ and a point $x_0 \in X$ whose orbit under the action of $G$ is dense. Let $(F_n)_{n \in \mathbb{N}}$ be a countable sequence of closed nowhere dense subsets of $X \times X$ with

$$E \subset \bigcup_{n \in \mathbb{N}} F_n.$$

The theorem is proved by a construction. We build a sequence of nonempty open sets $(V_s)_{s \in 2^{<N}}$ in $X$ and an array of group elements $\{(g_{s,t})_{s,t \in \{0,1\}^n} : n \in \mathbb{N}\}$ such that

(i) for $s : \{0, 1, \ldots, n-1\} \to \{0, 1\}$ we have

$$d(V_s) < \frac{1}{n};$$

(ii)

$$V_s \subset V_t$$

for $s \supset t$ and, moreover,

$$\overline{V_s} \subset V_t$$

if $t \subsetneq s$;

(iiia) for $s, t \in 2^n$ and $u \in 2^{<N}$ we have

$$g_{s,t} \cdot V_{s^\frown u} = V_{t^\frown u}; \text{ and}$$

(iiib) in fact if $\vec{0}$ is the sequence of all zeroes of length $n$ and $s, t \in 2^n$ then we have $g_{s,t} = g_{\vec{0},t} \circ g_{\vec{0},s}^{-1}$;

(iiic) for $s, t \in 2^n$ and $u \in 2^{<N}$ we have

$$g_{s,t} = g_{s^\frown u, t^\frown u};$$

(iv) for $s, t \in 2^n$ with $s \neq t$ we have

$$V_s \times V_t \cap \left( \bigcup_{m \leq n} F_m \right) = \emptyset.$$

If we succeed in doing all this, then the theorem follows quickly. We define for each $f \in 2^N$ a corresponding

$$x_f \in \bigcap_{n \in \mathbb{N}} V_{f|_n}.$$

There will be a unique such point $x_f$ in the infinite intersection by (i) and (ii); and then by the nature of the construction we have

$$f \mapsto x_f$$

as a Borel, in fact continuous, map. From (iiia) we have that if $N \in \mathbb{N}$ is the largest integer with $f_1(N-1) \neq f_2(N-1)$ then

$$g_{f_1|_N, f_2|_N} \cdot x_{f_1} = x_{f_2}$$

and thus

$$x_{f_1} E x_{f_2}.$$

Moreover (iv) yields that if there are infinitely many $n$ with $f_1(n) \neq f_2(n)$ then there are infinitely many $n$ with

$$\left( x_{f_1}, x_{f_2} \right) \notin \bigcup_{m \leq n} F_n,$$

and so

$$\left( x_{f_1}, x_{f_2} \right) \notin E \subset \bigcup_{n \in \mathbb{N}} F_n.$$

So let us now concentrate on showing that some such sequence of open sets and array of group elements has been produced. We do it by induction on $lh(s)$, and for this purpose assume that $(V_s)_{s \in 2^{\leq n}}$ and $\{(g_{s,t})_{s,t \in \{0,1\}^m} : m \leq n\}$ have the properties indicated above. We will begin by refining the open set $V_{\vec{0}}$, where $\vec{0}$ is the sequence of length $n$ which takes zero as its value on each coordinate.

Remember that the group $G$ is acting by homeomorphisms, and so

$$\{(x,y) : (g \cdot x, h \cdot y) \in F\}$$

is closed nowhere dense for any $g, h \in G$ and closed nowhere dense $F \subset X \times X$. Moreover, the finite union of closed nowhere dense sets is again closed nowhere dense.

Thus we may chose some open nonempty $W, W' \subset V_{\vec{0}}$ such that for all $g, h \in \{(g_{s,t})_{s,t \in \{0,1\}^m} : m \leq n\}$,

$$g \cdot W \times h \cdot W' \cap \left( \bigcup_{m \leq n+1} F_m \right) = \emptyset.$$

After further refinement we may assume that

$$d(g \cdot W), d(g \cdot W') < \frac{1}{n+1}$$

and

$$\overline{g_{\vec{0},s} \cdot W}, \overline{g_{\vec{0},s} \cdot W'} \subset V_s$$

for all $s \in 2^n$.

Then since $[x_0]_G$ is dense we may choose

$$x \in W \cap [x_0]_G,$$

$$x' \in W' \cap [x_0]_G.$$

In particular there will be some $h_0 \in G$ with

$$h_0 \cdot x = x'.$$

Since $G$ acts by homeomorphisms we may find $V_{\vec{0}^\frown 0} \subset W$, $V_{\vec{0}^\frown 1} \subset W'$ with

$$h_0 \cdot V_{\vec{0}^\frown 0} = V_{\vec{0}^\frown 1}.$$

At this stage we can finish up with

$$V_{s\smallfrown 0} = g_{\vec{0},s} \cdot V_{\vec{0}\smallfrown 0},$$

$$V_{s\smallfrown 1} = g_{\vec{0},s} h_0 \cdot V_{\vec{0}\smallfrown 0} = g_{\vec{0},s} \cdot V_{\vec{0}\smallfrown 1},$$

$$g_{\vec{0}\smallfrown 0,s\smallfrown 1} = g_{\vec{0},s} \circ h_0,$$

for all $s \in 2^n$. The decisions

$$g_{\vec{0}\smallfrown 0,s\smallfrown 0} = g_{\vec{0},s}$$

$$g_{s\smallfrown i, t\smallfrown i} = g_{s,t}$$

for all $s, t \in 2^n$, $i \in \{0, 1\}$ are forced on us by (iiib) and (iiic). Taking

$$g_{s\smallfrown 0, t\smallfrown 1} = g_{\vec{0},t} \circ h_0 \circ g_{\vec{0},s}^{-1}$$

and

$$g_{s\smallfrown 1, t\smallfrown 0} = g_{t\smallfrown 0, s\smallfrown 1}^{-1}$$

provides us with the full armory of group elements to go up to $n + 1$.   ⊣

COROLLARY 6.1. *Let $F \subset \mathcal{L}_{\omega_1,\omega}$ be a countable fragment, $T \subset F$ a complete theory. Then either*

$$E_0 \leq_B {\cong} |_{\mathrm{Mod}(T)}$$

*or there is an F-atomic model.*

PROOF. Completeness of the theory gives that every isomorphism type is dense in $(\mathrm{Mod}(T), \tau_F)$. The orbit equivalence relation is induced by a continuous action of the infinite symmetric group, $S_\infty$, and so we can use the previous lemma to conclude that either

$$E_0 \leq_B {\cong} |_{\mathrm{Mod}(T)}$$

or the orbit equivalence relation is not meager; and in this later case we saw in Section 4 that we must have an atomic model.   ⊣

DEFINITION 6.2. A topological group[16] is a *Polish group* if it is Polish as a topological space.

EXERCISES 6.3.

1. Let $G$ be a Polish group acting continuously on a Polish space $X$. Show that there is a Borel function

$$\theta : X \to \mathbb{R}$$

such that $\theta(x_1) = \theta(x_2)$ if and only if the orbits of $x_1$ and $x_2$ have the same closure.

---

[16]That is, a group equipped with a topology under which the group operations of multiplication and inversion are continuous.

2. Let $G$ be a locally compact Polish group acting continuously on a Polish space $X$ with orbit equivalence relation $E_G$.
  (a) Show that $E_G$ is $F_\sigma$ as a subset of $X \times X$.
  (b) Show that if there is dense orbit then it must contain an open set.
  (c) Show that in general, even without a dense orbit, we have either

$$E_G \leq \mathrm{id}(\mathbb{R})$$

or

$$E_0 \leq_B E_G.$$

## §7. Final theorem.

DEFINITION 7.1. HC denotes the collection of hereditarily countable sets. A formula $\psi(\vec{x})$ in the language of set theory is $\Sigma_0$ if the only quantifiers appearing are the bounded quantifiers

$$\forall x \in y$$
$$\exists x \in y.$$

A subset $A \subset \mathrm{HC}$ is said to be $\underset{\sim}{\Sigma}_1^{\mathrm{HC}}$ if there is some $\Sigma_0$ formula $\psi$ and parameter $a \in \mathrm{HC}$ such that $A$ equals the set of $b \in \mathrm{HC}$ for which

$$(\mathrm{HC}, \in) \models \exists x \psi(a, b, x).$$

$D \subset \mathrm{HC}$ is $\underset{\sim}{\Delta}_1^{\mathrm{HC}}$ if both it and its complement are $\underset{\sim}{\Sigma}_1^{\mathrm{HC}}$. I will say that a function is $\underset{\sim}{\Delta}_1^{\mathrm{HC}}$ if its graph and domain are both $\underset{\sim}{\Delta}_1^{\mathrm{HC}}$.

It will be convenient to phrase the next result in terms of $\underset{\sim}{\Delta}_1^{\mathrm{HC}}$ reductions; but I will skip over any detailed verifications about certain functions and sets appearing in the definition being $\underset{\sim}{\Delta}_1^{\mathrm{HC}}$. Instead here are some facts without proof. For many of these facts there are similar facts with similar proofs holding for the recursive functions.

FACTS 7.2.
1. Ordinal addition, multiplication, and exponentiation are $\underset{\sim}{\Delta}_1^{\mathrm{HC}}$.
2. If

$$G: \omega_1 \times \mathrm{HC} \to \mathrm{HC}$$
$$H: \mathrm{HC} \to \mathrm{HC}$$

are $\underset{\sim}{\Delta}_1^{\mathrm{HC}}$, then so is the function defined recursively by

$$F(\alpha + 1) = H(G(\alpha, F(\alpha)))$$

and

$$F(\lambda) = \bigcup \{F(\alpha) : \alpha < \lambda\}$$

when $\lambda$ a limit.

3. For $a \in HC$

$$\alpha \mapsto L_\alpha(a)$$

is $\underset{\sim}{\Delta}_1^{HC}$, where we here use $L_\alpha(a)$ to denote the $\alpha$th level of the constructible hierarchy built over $a$.

4. The $\underset{\sim}{\Delta}_1^{HC}$ recursive functions are closed under $\mu$-recursion with respect to the ordinals. Thus if for each $z \in HC$ there exists $\alpha < \omega_1$ with

$$L_\alpha(z, a) \models \psi(a, z),$$

then the function assigning to each $z$ the least such $\alpha$ is $\underset{\sim}{\Delta}_1^{HC}$.

5. Given $\mathcal{L}$ a countable language, there is a $\underset{\sim}{\Delta}_1^{HC}$ function assigning to each $\mathcal{M} \in \mathrm{Mod}(T)$ and countable fragment $F \subset \mathcal{L}_{\omega_1,\omega}$ the theory of $\mathcal{M}$ in that fragment.

6. Every *set recursive function* (in the sense of Garvin Melles) is $\underset{\sim}{\Delta}_1^{HC}$; and in fact the theorem below could equally be phrased in terms of set recursive functions instead of $\underset{\sim}{\Delta}_1^{HC}$.

THEOREM 7.1. *Let $\mathcal{L}$ be a countable language and $\sigma \in \mathcal{L}_{\omega_1,\omega}$. Then either*

(i) $E_0 \leq_B \cong |_{\mathrm{Mod}(\sigma)}$; *or*
(ii) *there is a $\underset{\sim}{\Delta}_1^{HC}$ function*

$$\theta: \mathrm{Mod}(\sigma) \to 2^{<\omega_1}$$

*such that for all $\mathcal{M}, \mathcal{N} \in \mathrm{Mod}(\sigma)$*

$$\mathcal{M} \cong \mathcal{N} \Leftrightarrow \theta(\mathcal{M}) = \theta(\mathcal{N}).$$

*In other words, $\cong |_{\mathrm{Mod}(\sigma)}$ is either as complicated as $E_0$ or allows a $\underset{\sim}{\Delta}_1^{HC}$ assignment of bounded subset of $\aleph_1$ as complete invariants.*

PROOF. Assume $E_0$ is not Borel reducible to $\cong |_{\mathrm{Mod}(\sigma)}$; by the corollary to Becker-Kechris this will ensure that we have many atomic models no matter which fragment we choose to consider.

First fix a fragment $F_0$ containing $\sigma$. For each $\alpha < \omega_1$ and positive integer $n$ let $P_{\alpha,n}$ be a new $n$-ary relation; we can do this so that

$$(\alpha, n) \mapsto P_{\alpha,n}$$

is a $\underset{\sim}{\Delta}_1^{HC}$ function. We let $\mathcal{L}^\alpha$ be the language generated by $\mathcal{L}$ and $\{P_{\beta,n}: n > 0, \beta < \alpha\}$ and we let $F_\alpha \subset \mathcal{L}^\alpha_{\omega_1,\omega}$ be the fragment generated by $F_0$ and all formulas of the form

$$\bigvee_{\{\beta: \gamma \leq \beta < \gamma'\}} P_{\beta,n}(\vec{x})$$

for $\gamma < \gamma' \leq \alpha$. We may simultaneously fix a $\underset{\sim}{\Delta}_1^{HC}$ increasing sequence of ordinals,

$$(\gamma_\alpha)_{\alpha \in \omega_1},$$

and bijections

$$\pi_\alpha \colon \gamma_\alpha \to F_\alpha$$

which are uniformly $\underset{\sim}{\Delta}_1^{HC}$ in the sense that

$$(\alpha, \beta) \mapsto \pi_\alpha(\beta)$$

is a $\underset{\sim}{\Delta}_1^{HC}$ function.

Fix $\mathcal{M} \in \mathrm{Mod}(\sigma)$. We describe a a sequence of expansions

$$\mathcal{M}_0 = \mathcal{M}, \mathcal{M}_1, \ldots, \mathcal{M}_\alpha, \ldots$$

out through the ordinals, the isomorphism type of each $\mathcal{M}_\alpha$ depending only on the isomorphism type of $\mathcal{M}$ and each $\mathcal{M}_\alpha$ an $\mathcal{L}^\alpha$-structure; at each $\alpha$ we let $T_\alpha^{\mathcal{M}}$ be the theory of $\mathcal{M}_\alpha$ with respect to the fragment $F_\alpha$.

Thus in particular we begin with

$$T_0^{\mathcal{M}} = \mathrm{Th}_F(\mathcal{M}),$$

the theory of $\mathcal{M}$ with respect to $F$.

CLAIM. $T_0^{\mathcal{M}}$ has an atomic model.

PROOF OF CLAIM. Or else Becker-Kechris gives

$$E_0 \leq_B \cong |_{\mathrm{Mod}(T_0^{\mathcal{M}})},$$

$$E_0 \leq_B \cong |_{\mathrm{Mod}(\sigma)}. \qquad\qquad \text{CLAIM} \dashv$$

We then define $\mathcal{M}_1$ to be the expansion of $\mathcal{M}$ to $\mathcal{L}^1$ as follows. For each $a_1, a_2, \ldots, a_n \in \mathcal{M}$ we let

$$\mathcal{M}_1 \models P_{0,n}(\vec{a})$$

if and only if $\vec{a}$ realizes a principal type relative to the theory $T_{\mathcal{M}}^0$. Then we let $T_1^{\mathcal{M}}$ be the theory of $\mathcal{M}_1$ relative to the fragment $F_1$.

And we can keep going in the obvious way. We let

$$T_\alpha^{\mathcal{M}} = \mathrm{Th}_{F_\alpha}(\mathcal{M}_\alpha);$$

$\mathcal{M}_{\alpha+1}$ is the expansion to $\mathcal{L}^{\alpha+1}$ obtained by letting

$$\mathcal{M}_{\alpha+1} \models P_{\alpha,n}(\vec{a})$$

if $lh(\vec{a}) = n$ and $\vec{a}$ realizes a principal type over the theory $T_\alpha^{\mathcal{M}}$. At limit stages we can let $\mathcal{M}_\lambda$ be the unique model in the language

$$\mathcal{L}^\lambda = \bigcup_{\alpha < \lambda} \mathcal{L}^\alpha$$

which expands all the preceding

$$\{\mathcal{M}_\alpha \colon \alpha < \lambda\}.$$

One possibility is that regardless of our particular choice of $\mathcal{M}$ we always arrive at some $F_\kappa$ with $\mathcal{M}_\kappa$ the $T_\kappa^{\mathcal{M}}$-atomic model for $F_\kappa$. Then we are done. We can just let $\kappa(\mathcal{M})$ be the first such $\kappa$ for $\mathcal{M}$ and let

$$\theta(\mathcal{M}) = \{\beta < \gamma_{\kappa(\mathcal{M})} : \mathcal{M}_{\kappa(\mathcal{M})} \models \pi_{\kappa(\mathcal{M})}(\beta)\} \cup \{\gamma_{\kappa(\mathcal{M})}\}$$

to obtain an invariant which indicates to us $\gamma_{\kappa(\mathcal{M})}$, and hence $\kappa(\mathcal{M})$, and hence the fragment $F_{\kappa(\mathcal{M})}$, as well as the theory $T_{\kappa(\mathcal{M})}^{\mathcal{M}}$ and the fact that $\mathcal{M}_{\kappa(\mathcal{M})}$ is atomic with respect to this theory; since each complete countable theory has at most one atomic model up to isomorphism, the invariant indicates the isomorphism type of $\mathcal{M}_{\kappa(\mathcal{M})}$; since $\mathcal{M}_{\kappa(\mathcal{M})}$ is up to isomorphism an invariant of $\mathcal{M}$, we obtain that $\theta(\mathcal{M})$ is a complete invariant for the isomorphism type of $\mathcal{M}$. And finally in this case it would be routine to establish that the association

$$\mathcal{M} \mapsto \theta(\mathcal{M})$$

is $\underset{\sim}{\Delta}_1^{HC}$.

So instead let us try to show that some such ordinal $\kappa(\mathcal{M})$ always exists. And for this let us assume that $\mathcal{M}$ is a counterexample and try to derive a contradiction.

CLAIM. For each $n > 0$ and $\delta < \omega_1$ there is a larger countable ordinal $\delta' > \delta$ such that for each $a_1, a_2, \ldots, a_n \in \mathcal{M}$

$$\mathcal{M}_{\delta'} \models \bigvee_{\alpha \in [\delta, \delta')} P_{\alpha, n}(\vec{a}).$$

PROOF OF CLAIM. Instead we obtain that each $\delta'$

$$\mathcal{M}_{\delta'} \models \exists \vec{x} \bigwedge_{\alpha \in [\delta, \delta')} \neg P_{\alpha, n}(\vec{x}).$$

Since $T_\delta^{\mathcal{M}}$ has an atomic model, by Becker-Kechris, we obtain some principal type refining the type

$$\bigwedge_{\alpha \in [\delta, \delta')} \neg P_{\alpha, n}(\vec{x}),$$

and thus by the nature of our construction some $\vec{a}^{\delta'}$ with

$$\mathcal{M}_{\delta'+1} \models P_{\delta'}\left(\vec{a}^{\delta'}\right) \bigwedge_{\alpha \in [\delta, \delta')} \neg P_{\alpha, n}\left(\vec{a}^{\delta'}\right);$$

but then

$$\delta' \mapsto \vec{a}^{\delta'}$$

gives us $\aleph_1$ distinct $n$-tuples in $\mathcal{M}$, with an obvious contradiction to its countability.                                                            CLAIM $\dashv$

Thus by repeated application of this claim we can find an increasing sequence of ordinals,

$$(\delta_\alpha)_{\alpha<\omega_1},$$

such that for each $\alpha \in \omega_1$, $n > 0$, $\vec{a} \in \mathcal{M}^n$ there is some

$$\beta > \bigcup_{\alpha'<\alpha} \delta_{\alpha'}$$

with

$$\beta \leq \delta_\alpha$$

and

$$\mathcal{M}_{\delta_\alpha} \models P_{\beta,n}(\vec{a}).$$

We may also fix for each $\beta < \omega_1$ and $n > 0$ a sequence of formulas $(\psi^m_{\beta,n})_m$ such that

$$\mathcal{M}_{\beta+1} \models \forall x_1, x_2, \ldots, x_n \left( P_{\beta,n}(\vec{x}) \Leftrightarrow \bigvee_{m\in\mathbb{N}} \psi^m_{\beta,n}(\vec{x}) \right)$$

and each $\psi^m_{\beta,n}(\vec{x}) \in F_\beta$ defines a principal type over $T^\mathcal{M}_\beta$. Unfortunately the disjunction $\bigvee_{m\in\mathbb{N}} \psi^m_{\beta,n}(\vec{x})$ has not been placed in any of our fragments, so we need to observe that the equivalence

$$\mathcal{N} \models \forall x_1, x_2, \ldots, x_n \left( P_{\beta,n}(\vec{x}) \Leftrightarrow \bigvee_{m\in\mathbb{N}} \psi^m_{\beta,n}(\vec{x}) \right)$$

holds for any sufficiently "generic" $\mathcal{N}$ in the space of $T^\mathcal{M}_\beta$ models, for any $\alpha > \beta$.

CLAIM. For each $n > 0$, $\vec{a} \in \mathbb{N}^n$ and $\beta < \alpha$, the set of

$$\mathcal{N} \in \left( \text{Mod}\left( T^\mathcal{M}_\alpha \right), \tau_{F_\alpha^\mathcal{M}} \right)$$

with

$$\mathcal{N} \models P_{\beta,n}(\vec{a}) \Leftrightarrow \bigvee_{m\in\mathbb{N}} \psi^m_{\beta,n}(\vec{a})$$

is open dense.

PROOF OF CLAIM. It is a straight calculation to determine that the set of such $\mathcal{N}$ is open. For density we use that the isomorphism type of $\mathcal{M}_\alpha$ is dense in $(\text{Mod}(T^\mathcal{M}_\alpha), \tau_{F_\alpha^\mathcal{M}})$ and

$$\mathcal{M}_{\beta+1} \models \forall x_1, x_2, \ldots, x_n \left( P_{\beta,n}(\vec{x}) \Leftrightarrow \bigvee_{m\in\mathbb{N}} \psi^m_{\beta,n}(\vec{x}) \right).$$

CLAIM $\dashv$

Thus for each $\alpha$ there is a comeager set

$$C_\alpha \subset \left(\mathrm{Mod}\left(T_\alpha^{\mathcal{M}}\right), \tau_{F_\alpha^{\mathcal{M}}}\right)$$

such that for all $\beta < \alpha, n > 0, \mathcal{N} \in C_\alpha$,

$$\mathcal{N} \models \forall x_1, x_2, \ldots, x_n \left(P_{\beta,n}(\vec{x}) \Leftrightarrow \bigvee_{m \in \mathbb{N}} \psi_{\beta,n}^m(\vec{x})\right).$$

In the next claim I will use $\mathcal{N}|_{\mathcal{L}}$ to indicate the reduction of some $\mathcal{L}^\alpha$ model to our base language $\mathcal{L}$. As usual,

$$\varphi_\kappa^{\vec{\ell}, \mathcal{N}|_{\mathcal{L}}}$$

will indicate the $\kappa$th approximation to the Scott sentence of $\vec{\ell}$ over $\mathcal{N}|_{\mathcal{L}}$.

CLAIM. For each $\mathcal{N} \in C_\alpha, \delta_\kappa \leq \beta < \alpha < \omega_1, n > 0, m \in \mathbb{N}, \vec{\ell}, \vec{k} \in \mathbb{N}^n$ we have that if

$$\mathcal{M}_\alpha \models \psi_{\beta,n}^m(\vec{k})$$

and

$$\mathcal{N} \models \psi_{\beta,n}^m(\vec{\ell})$$

then

$$\varphi_\kappa^{\vec{\ell}, \mathcal{N}|_{\mathcal{L}}} = \varphi_\kappa^{\vec{k}, \mathcal{M}}.$$

PROOF OF CLAIM. We prove this by induction on $\kappa$. It should be clear for $\kappa = 0$, since $\psi_{\beta,n}^m \in F_0$ defines a principal type over $F_0$, and thus is sufficient to decide the quantifier-free type, even if $\mathcal{N}$ does not belong to our comeager set $C_\alpha$. The limit steps follow almost vacuously from the structure of the Scott analysis and its requirement that we take conjunctions at limit stages.

For successor steps, let us suppose $\delta_{\kappa+1} \leq \beta < \alpha < \omega_1$,

$$\mathcal{M}_\alpha \models \psi_{\beta,n}^m(\vec{k}),$$

and

$$\mathcal{N} \models \psi_{\beta,n}^m(\vec{\ell}).$$

Choose some $\vec{a} \in \mathcal{N}$; we need to show that there is some other $\vec{b} \in \mathcal{M}$ with

$$\varphi_\kappa^{\vec{\ell}\vec{a}, \mathcal{N}|_{\mathcal{L}}} = \varphi_\kappa^{\vec{k}\vec{b}, \mathcal{M}}.$$

But since $\mathcal{N}$ and $\mathcal{M}_\beta$ realize the same theory it must be the case that

$$\mathcal{N} \models \forall x_1, \ldots, x_n \bigvee_{\{\beta' : \gamma_\kappa \leq \beta' < \gamma_{\kappa+1}\}} P_{\beta',n}(\vec{x}),$$

and thus, in particular, for some $\beta' \in [\gamma_\kappa, \gamma_{\kappa+1})$ and $n' = lh(\vec{\ell}) + lh(\vec{a})$ we have

$$\mathcal{N} \models P_{\beta',n'}(\vec{\ell}, \vec{a});$$

then by assumption on $\mathcal{N} \in C_\alpha$ we have some $m'$ with

$$\mathcal{N} \models \psi^{m'}_{\beta',n'}(\vec{\ell}, \vec{a});$$

then since $\psi^m_{\beta,n}$ decides the type of $\vec{k}$ over the fragment $F_{\beta'}$ we have

$$\mathcal{M}_\alpha \models \exists \vec{y}\, \psi^{m'}_{\beta',n'}(\vec{k}, \vec{y});$$

and so for some $\vec{b}$

$$\mathcal{M}_\alpha \models \psi^{m'}_{\beta',n'}(\vec{k}, \vec{b})$$

and so by the inductive assumption,

$$\varphi_\kappa^{\vec{\ell}\vec{a},\mathcal{N}|_{\mathcal{L}}} = \varphi_\kappa^{\vec{k}\vec{b},\mathcal{M}}.$$

The converse direction is that for all $\vec{b} \in \mathcal{M}$ there is $\vec{a} \in \mathcal{N}$ with $\varphi_\kappa^{\vec{\ell}\vec{a},\mathcal{N}|_{\mathcal{L}}} = \varphi_\kappa^{\vec{k}\vec{b},\mathcal{M}}$. This is similar, but easier.                                        CLAIM ⊣

Thus we obtain that for any $\alpha > \alpha(\mathcal{M}) + 2$ and $\mathcal{N} \in C_\alpha$

$$\varphi_{\alpha(\mathcal{M})+2}^{\emptyset,\mathcal{N}|_{\mathcal{L}}} = \varphi_{\alpha(\mathcal{M})+2}^{\emptyset,\mathcal{M}},$$

and thus by Scott's theorem

$$\mathcal{N}|_{\mathcal{L}} \cong \mathcal{M};$$

but then it follows from the definition of the various $\mathcal{M}_\beta$s that the isomorphism from $\mathcal{N}|_{\mathcal{L}}$ to $\mathcal{M}$ lifts to one from $\mathcal{N}$ to $\mathcal{M}_\beta$ for each $\beta < \alpha$.

Thus the isomorphism type of $\mathcal{M}_\alpha$ *will* be comeager in some

$$\left(\mathrm{Mod}\left(T_\alpha^\mathcal{M}\right), \tau_{F_\alpha^\mathcal{M}}\right);$$

and thus $\mathcal{M}_\alpha$ *will* be atomic, and so the process must have terminated at some stage after all.                                                                        ⊣

In general this theorem can be improved by slightly sharpening the reduction to one that is *provably* or *absolutely* $\underset{\sim}{\Delta}_1^{HC}$; these complexity classes are a little technical to define,[17] but have the advantage of being just inside the region for which ZFC alone can prove regularity properties—such as any *absolutely* $\underset{\sim}{\Delta}_1^{HC}$ function between Polish spaces is continuous on a comeager set. In this sharper form the theorem would then become a dichotomy theorem: (i) and a suitably amended version of (ii) would be incompatible.

A respect in which the theorem cannot be sharpened is with regard to the *kinds of* invariants we obtain in (ii). For instance we cannot ask that we have, say, elements of $2^\mathbb{N}$ being assigned as complete invariants. The isomorphism relation on abelian $p$-groups provides a counterexample under suitable set theoretical assumptions for general $\underset{\sim}{\Delta}_1^{HC}$ functions and outright in ZFC for the absolutely $\underset{\sim}{\Delta}_1^{HC}$.

---

[17]See, e.g., *Classification and orbit equivalence relations*, G. Hjorth, AMS, Rhode Island, 2000, for the precise definitions.

However, this class is not first-order definable, so I don't know whether one might hope to prove that for any first-order theory $T$ we have either

(i) $E_0 \leq_B \cong |_{\text{Mod}(T)}$, or

(ii') $\cong |_{\text{Mod}(T)} \leq_B \text{id}(2^{\mathbb{N}})$.

Even showing this with (ii') replaced by

(ii'') there is a $\underset{\sim}{\Delta}_1^{HC}$ assignment of elements of $2^{\mathbb{N}}$ as complete invariants,

would be extremely interesting, and sufficient to prove Vaught's conjecture under large cardinal assumptions, or even prove Vaught's conjecture outright in ZFC if, as is most likely, one could obtain

(ii*) there is an absolutely $\underset{\sim}{\Delta}_1^{HC}$ assignment of elements of $2^{\mathbb{N}}$ as complete invariants.

§8. **More reading.** There is always more to read. In this case especially there are a number of issues we only touched on which could have been discussed at length.

The basic references for the descriptive set theory of group actions connectioned with the isomorphism relation on countable models are Becker and Kechris [1996] and Sami [1994].

The serious mathematical discussion of dichotomy theorems for Borel equivalence relations was initiated in Harrington, Kechris, and Louveau [1990].

A more recent survey is given by Hjorth and Kechris [1997].

A continued discussion of some of the material above, in much the same elementary style, is given by Hjorth [2000].

This includes some easy proofs of things we didn't quite get to, such as $F_2$ not being Borel reducible to any of the countable Borel equivalence relations, such as $E_0$.

The main theorem, which we finished with, was first presented in Hjorth and Kechris [1995] and is also independently due to Howard Becker. The proof given there is rather different, in that it is derived as a corollary of Harrington-Kechris-Louveau; and indeed in this paper the main battle was to extend the result to general $\underset{\sim}{\Sigma}_1^1$ equivalence relations, while above we have been primarily concerned with developing an argument that uses no nontrivial descriptive set theory. If there had been more time we might have discussed special versions of this dichotomy theorem which can be proved for specific classes of Polish groups. Then one often can hope to replace (ii) by (ii') $\cong |_{\text{Mod}(T)} \leq_B \text{id}(2^{\mathbb{N}})$.

In Hjorth and Solecki [1999] this was done for the orbit equivalence relations induced by continuous actions of nilpotent and invariantly metrizable Polish groups, while Becker [1998] obtained more general results for solvable or under the assumption of a complete left invariant metric.

A rather different approach to some of these questions can be found in Melles [1992].

### REFERENCES

HOWARD BECKER [1998], *Polish group actions: Dichotomies and generalized embeddings*, **Journal of the American Mathematical Society**, vol. 11, pp. 397–449.

HOWARD BECKER AND ALEXANDER S. KECHRIS [1996], **The descriptive set theory of Polish group actions**, Cambridge University Press, Cambridge.

L. A. HARRINGTON, A. S. KECHRIS, AND A. LOUVEAU [1990], *A Glimm-Effros dichotomy for Borel equivalence relations*, **Journal of the American Mathematical Society**, vol. 3, no. 4, pp. 903–928.

GREG HJORTH [2000], **Classification and orbit equivalence relations**, American Mathematical Society, Providence.

GREG HJORTH AND ALEXANDER S. KECHRIS [1995], *Analytic equivalence relations and Ulm-type classifications*, **The Journal of Symbolic Logic**, vol. 60, no. 4, pp. 1273–1300.

GREG HJORTH AND ALEXANDER S. KECHRIS [1997], *New dichotomies for Borel equivalence relations*, **The Bulletin of Symbolic Logic**, vol. 3, no. 3, pp. 329–346.

GREG HJORTH AND SLAWOMIR SOLECKI [1999], *Vaught's conjecture and the Glimm-Effros property for Polish transformation groups*, **Transactions of the American Mathematical Society**, vol. 351, no. 7, pp. 2623–2641.

GARVIN MELLES [1992], *One cannot show from ZFC that there is an Ulm-type classification of the countable torsion-free abelian groups*, **Set theory of the continuum (Berkeley, CA, 1989)**, Springer, New York, pp. 293–309.

RAMEZ L. SAMI [1994], *Polish group actions and the Vaught conjecture*, **Transactions of the American Mathematical Society**, vol. 341, no. 1, pp. 335–353.

DEPARTMENT OF MATHEMATICS
  UCLA
    LOS ANGELES CA 90095-1555, USA
*E-mail*: greg@math.ucla.edu

# Index for *Countable models and the theory of Borel equivalence relations*

# MODEL THEORY OF DIFFERENCE FIELDS

ZOÉ CHATZIDAKIS

**§1. Introduction.** This is the set of notes from a special topics course I gave at Notre Dame in Fall 2000 (September 20–October 13). I develop here some of the model theory of difference fields (a difference field is simply a field with a distinguished automorphism $\sigma$) at a fairly elementary level. I reproduce many of the classical proofs of stability theory in the particular context we are working in, my feeling being that a proof in a concrete situation is much easier to understand than in a more general context. I tried to avoid using results from stability theory and succeeded except at one or two places (where the neophyte is asked to just accept the result). I also inserted some comments for people with a working knowledge of stability theory, and these are enclosed by the symbols □□□; these comments can be skipped.

The notes are organized as follows. Section 1 gives some preliminary algebraic results and definitions (difference fields, varieties, Zariski topology, etc.). Section 2 introduces the theory ACFA of generic difference fields and proves elementary results about it. Section 3 introduces the notions of independence and SU-rank and shows various results about them. In Section 4 we study the fixed field, and in Section 5 the notions of orthogonality and modularity. Section 6 introduces generics and stabilizers of groups. Finally, Section 7 contains some of the hard results in the area and the applications by Hrushovski to problems in number theory. At the end of the notes, you will find some references and "further reading" on difference fields, with comments.

**Notation.**

| | |
|---|---|
| $\mathbb{N}, \mathbb{Z}, \mathbb{Q}, \mathbb{C}$ | the natural numbers, the integers, the rational and complex numbers |
| $\mathcal{L}(E)$ | language obtained by adjoining to $\mathcal{L}$ constant symbols for elements of $E$ |

I would like to thank the Notre Dame logicians for their hospitality and for the opportunity of giving these lectures. Special thanks to A. Berenstein for his careful reading and many helpful suggestions.

**The Notre Dame Lectures**
Edited by P. Cholak
Lecture Notes in Logic, 18

| | |
|---|---|
| $\mathbb{F}_q$ | the field with $q$ elements |
| $AB$ | subfield of $\Omega$ generated by $A$ and $B$ |
| $A[B]$ | subring of $\Omega$ generated by $A$ and $B$ |
| $A^{\mathrm{alg}}$ | algebraic closure of the field $A$ |
| $K[X_1, \ldots, X_n]_\sigma$ | $= K[X_1, \ldots, X_n, X_1^\sigma, \ldots, X_n^\sigma, \ldots, X_i^{\sigma^j}, \ldots]$ |
| | $=$ difference polynomial ring in $X_1, \ldots, X_n$ |
| $I(S)$ | ideal of polynomials vanishing on the set $S$ |
| $I_\sigma(S)$ | ideal of difference polynomials vanishing on the set $S$ |
| $I(\bar{a}/K)$ | ideal of polynomials over $K$ vanishing at $\bar{a}$ |
| $I_\sigma(\bar{a}/K)$ | ideal of difference polynomials over $K$ vanishing at $\bar{a}$ |
| $\mathrm{cl}_\sigma(A)$ | smallest difference field containing $A$ |
| | $=$ field generated by $\{\sigma^i(A) \mid i \in \mathbb{Z}\}$, with the action of $\sigma$ |
| $\mathrm{acl}_\sigma(A)$ | $\mathrm{cl}_\sigma(A)^{\mathrm{alg}}$ |
| $U^\sigma$ | variety conjugate of $U$ under $\sigma$ |
| $U(K)$ | points of the algebraic set $U$ with their coordinates in $K$ |
| $\mathrm{qfdiag}(E)$ | set of quantifier-free $\mathcal{L}(E)$-sentences which hold in some $\mathcal{L}$-structure containing $E$ |
| $\mathrm{qftp}$ | quantifier-free type |
| $\mathrm{tp}_{\mathrm{ACF}}$ | type in the reduct to the language of fields $\{+, -, \cdot, 0, 1\}$ |
| $\mathrm{tr.deg}$ | transcendence degree |
| $\dim$ | dimension of an algebraic set |
| $\perp$ | orthogonal |
| $\mathrm{Tor}(A)$ | torsion points of the group $A$ |
| $\mathrm{Tor}_{p'}(A)$ | torsion points of the group $A$ of order prime to $p$ |
| □□□ | The text enclosed by these symbols is meant as comments for those with a working knowledge of stability theory. |

## §2. Definitions, preliminary results.

**2.1. Definition.** A *difference field* is a field $K$ with a distinguished endomorphism $\sigma$. An *inversive difference field* is a difference field where the endomorphism is onto (and therefore an automorphism). Given a difference field $K$, there is a "smallest" inversive difference field extending $K$, which is unique up to $K$-isomorphism. Therefore we will adopt the following convention:

*In what follows, all difference fields will be assumed to be inversive.*

Consider the language $\mathcal{L} = \{+, -, \cdot, 0, 1, \sigma\}$, where $+, -, \cdot$ are binary operations, $0, 1$ are constants, and $\sigma$ is a unary function symbol. Then every difference field is naturally an $\mathcal{L}$-structure. Inclusion of difference fields corresponds to inclusion of $\mathcal{L}$-structures, and similarly for morphisms of difference fields. A good reference for basic results on difference fields is R. M. Cohn's *Difference algebra* [1].

**2.2. Difference polynomials, difference equations.** Let $(K, \sigma)$ be a difference field, and $\bar{X} = (X_1, \ldots, X_n)$ a tuple of indeterminates. The difference polynomial ring in $\bar{X}$ over $K$, denoted $K[\bar{X}]_\sigma$, is defined as follows.

As a ring, $K[\bar{X}]_\sigma$ is simply the polynomial ring in the indeterminates $X_1, \ldots, X_n, X_1^\sigma, \ldots, X_n^\sigma, \ldots, X_1^{\sigma^m}, \ldots$.

The action of $\sigma$ on $K[\bar{X}]_\sigma$ is the one suggested by the names of the indeterminates. Then $K[\bar{X}]_\sigma$ is also an $\mathcal{L}$-structure.

REMARKS. (1) A *difference ring* is a ring with a distinguished injective endomorphism. So, $K[\bar{X}]_\sigma$ is a difference ring.

(2) This map $\sigma$ is injective, but is not surjective. One could instead take $K[X_j^{\sigma^i}]_{j=1,\ldots,n, i \in \mathbb{Z}}$.

The elements of $K[\bar{X}]_\sigma$ are called *difference polynomials*. If $f(\bar{X}) \in K[\bar{X}]_\sigma$, then $f(\bar{X}) = 0$ is called a *$\sigma$-equation*, and the set $\{\bar{a} \in K^n \mid f(\bar{a}) = 0\}$ is called a *$\sigma$-closed set*.

**2.3. Definitions.** (1) Let $T$ be a theory, $M$ a model of $T$. Then $M$ is *existentially closed* (among models of $T$) if whenever $N$ is a model of $T$ containing $M$, then every existential sentence with parameters in $M$ which is satisfied in $N$ is already satisfied in $M$.

(2) A *generic difference field* is a difference field $K$ such that every finite system of $\sigma$-equations (over $K$) which has a solution in some difference field extending $K$, has already a solution in $K$.

**2.4. Exercise 1.** Show that if $K$ is a generic difference field, then $K$ is an existentially closed difference field. (Hint: recall that in a field $x \neq 0 \iff \exists y \, xy = 1$.)

**2.5.** Our first goal is to show that the generic difference fields form an elementary class (with axiomatization called ACFA). We will then study some easy properties of this theory. We first need, however, to recall some definitions of basic algebraic geometry and of field theory. Proofs and details can be found in S. Lang's *Introduction to algebraic geometry* [2].

**2.6. Algebraic sets, varieties.** Here and in what follows we fix a large algebraically closed field $\Omega$. All fields considered will be subfields of $\Omega$. If $K$ is a subfield of $\Omega$, we denote by $K^{\mathrm{alg}}$ the algebraic closure of $K$, i.e., the set of elements of $\Omega$ which are algebraic over $K$. We first define affine algebraic sets.

*Affine n-space*, $\mathbb{A}^n(\Omega)$, is simply $\Omega^n$ (for $n \in \mathbb{N}$). An *algebraic subset* of $\Omega^n$ is a subset defined by polynomial equations $f_1(\bar{X}) = \cdots = f_m(\bar{X}) = 0$, where the $f_i(\bar{X})$ are polynomials in $\bar{X} = (X_1, \ldots, X_n)$, with coefficients in $\Omega$.

These sets are also called *Zariski closed*. The Zariski closed subsets of $\Omega^n$ generate a topology on $\Omega^n$ called the *Zariski topology*. This topology is Noetherian (the ascending chain condition on ideals of $\Omega[\bar{X}]$ implies the descending chain condition on Zariski closed sets).

**2.7.** Let $V$ be an algebraic set. We define

$$I(V) = \{f(\bar{X}) \in \Omega[\bar{X}] \mid f(\bar{a}) = 0 \text{ for all } \bar{a} \in V\}.$$

The coordinate ring of $V$, $\Omega[V]$, is defined to be $\Omega[\bar{X}]/I(V)$.

We say that $V$ is *defined over the subfield $K$* of $\Omega$, if $I(V)$ is generated by $I(V) \cap K[\bar{X}]$. In this case we can also form the coordinate ring of $V$ over $K$, $K[V] = K[\bar{X}]/I(V) \cap K[\bar{X}]$.

A closed subset $V$ of $\Omega^n$ is *irreducible* if whenever $V = V_1 \cup V_2$ with $V_1$, $V_2$ closed subsets of $\Omega^n$, then $V = V_1$ or $V = V_2$. An irreducible closed set is also called a *variety*.

If $V$ is a variety, we define the field of rational functions of $V$ to be the field of fractions of $\Omega[V]$; we denote it by $\Omega(V)$.

Because the Zariski topology is Noetherian, it follows that every algebraic set $V$ can be written (uniquely up to permutation) as

$$V = V_1 \cup \cdots \cup V_m$$

for some $m$ and varieties $V_1, \ldots, V_m$ such that $V_i \not\subseteq V_j$ for $i \neq j$. The $V_i$ are called the *irreducible components* of $V$.

**2.8. Some classical results.** (1) An (affine) algebraic set $V$ is a variety if and only if $I(V)$ is prime.

(2) If $V \subseteq W$ are algebraic sets, then $I(V) \supseteq I(W)$.

(3) (Nullstellensatz) Let $I$ be an ideal of $\Omega[\bar{X}]$, and let $V(I) = \{\bar{a} \in \Omega^n \mid f(\bar{a}) = 0 \text{ for all } f(\bar{X}) \in I\}$. Then

$$V(I) \neq \emptyset \iff I \neq (1).$$

(4) If $I \subseteq J$ are ideals in $\Omega[\bar{X}]$ then $V(I) \supseteq V(J)$.

(5) If $I$ is a *radical ideal* of $\Omega[\bar{X}]$ (i.e., $a^n \in I$ implies $a \in I$), then $I(V(I)) = I$.

(6) If $V$ is an algebraic subset of $\Omega^n$ then $V(I(V)) = V$.

(7) If $V_1, \ldots, V_m$ are the irreducible components of $V$, then $I(V_1), \ldots, I(V_m)$ are the minimal prime ideals containing $I(V)$, and $I(V) = \bigcap_{i=1,\ldots,m} I(V_i)$.

**2.9. Remarks.** (1) If $K$ is a subfield of $\Omega$, one can also define the notion of $K$-irreducible subsets of $\Omega^n$: an algebraic set $V$ is $K$-irreducible if it is defined over $K$ and cannot be written as the union of two proper algebraic subsets which are defined over $K$. This corresponds to the ideal $I(V) \cap K[\bar{X}]$ being prime.

(2) One can show that if the algebraic set $V$ is defined over $K$, then $V$ is a variety if and only if $V$ is $K^{\mathrm{alg}}$-irreducible (if and only if $I(V) \cap K[\bar{X}]$ generates a prime ideal in $K^{\mathrm{alg}}[\bar{X}]$).

(3) It may be that an algebraic set is defined by polynomial equations with their coefficients in $K$, but is not defined over $K$. This can happen in positive characteristic. Indeed, assume that $K$ is a field of characteristic $p > 0$, and

that $a \in K$ does not have a $p$th root in $K$. Then the set $V = \{a^{1/p}\}$ is defined by the equation $X^p - a = 0$ (with coefficients in $K$), but $I(V)$ is generated by $X - a^{1/p}$, and $V$ is therefore not defined over $K$ as $a^{1/p} \notin K$.

(4) If $V$ is an algebraic set, there is a smallest field $K$ over which $V$ is defined, and this field is called the *field of definition* of $V$. If $V$ is defined by polynomial equations with coefficients in $K$, then the field of definition of $V$ is contained in

— $K$ if char$(K) = 0$,

— the perfect hull of $K$ ($=$ closure of $K$ under $p$th roots) if char$(K) = p > 0$.

(Recall that in characteristic $p > 0$, the Frobenius map $x \mapsto x^p$ defines an injective endomorphism of the field $K$, as $(x + y)^p = x^p + y^p$. Hence, $K^{1/p^n} = \{a \in \Omega \mid a^{p^n} \in K\}$ is a subfield of $\Omega$, and so is the *perfect hull* $K^{1/p^\infty}$ of $K$: $K^{1/p^\infty} = \bigcup_{n \in \mathbb{N}} K^{1/p^n}$.)

**2.10. Dimension, generics.** Let $V \subseteq \Omega^n$ be a variety defined over $K$. We define the *dimension of* $V$, dim$(V)$, to be tr.deg$(\Omega[V]/\Omega)$. Note that if $V \subseteq \Omega^n$, then dim$(V) \leq n$. The tuple $\bar{a} = (a_1, \ldots, a_n)$ is a *generic* of $V$ over $K$ if the $K$-morphism $K[\bar{X}] \to K[\bar{a}]$ which sends $X_i$ to $a_i$ for $i = 1, \ldots, n$ has kernel $I(V) \cap K[\bar{X}]$.

If $V$ is an algebraic set, then dim$(V)$ is the maximum of the dimensions of the irreducible components of $V$.

REMARK. If $V$ is $K$-irreducible, then the irreducible components of $V$ are conjugate under Aut$(K^{\text{alg}}/K)$, and therefore have the same dimension. Thus one can also define dim$(V)$ as tr.deg$(K[V]/K)$ and define a notion of generic over $K$.

**2.11. Locus of a point.** Let $K$ be a subfield of $\Omega$ and $\bar{a} = (a_1, \ldots, a_n)$. Define

$$I(\bar{a}/K) = \{f(\bar{X}) \in K[\bar{X}] \mid f(\bar{a}) = 0\},$$

where $X = (X_1, \ldots, X_n)$. Then $I(\bar{a}/K)$ is a prime ideal of $K[\bar{X}]$. The set $V$ of points of $\Omega^n$ at which all elements of $I(\bar{a}/K)$ vanish, is therefore $K$-irreducible and is called the *locus* of $\bar{a}$ over $K$. Then

— $\bar{a}$ is a generic of $V$ over $K$.

— Let $\bar{b} \in \Omega^n$. Then $\bar{b} \in V \iff I(\bar{b}/K) \supseteq I(\bar{a}/K)$.

— Assume that $\bar{b} = (b_1, \ldots, b_n) \in V$. Then there is a unique $K$-morphism $\varphi \colon K[\bar{a}] \to K[\bar{b}]$ which sends $a_i$ to $b_i$ for $i = 1, \ldots, n$. This morphism is an isomorphism if and only if $\bar{b}$ is a generic of $V$.

— Let $V_1$ be an irreducible component of $V$, let $L = K(\bar{a}) \cap K^{\text{alg}}$, and let $\hat{L}$ be the normal closure of $L$ over $K$ ($=$ the field generated by all conjugates of $L$ under the action of Aut$(K^{\text{alg}}/K)$). Then the set $\{V_1^\tau \mid \tau \in \text{Aut}(\hat{L}/K)\}$ is precisely the set of irreducible components of $V$.

REMARK. Warning: if the characteristic is $p > 0$, the locus $V$ of $\bar{a}$ over $K$ is not necessarily *defined over K*. One has: $V$ is defined over $K$ if and only if $K(\bar{a})$ is a *separable extension of K*, i.e., $K(\bar{a})$ and $K^{1/p}$ are linearly disjoint over $K$.

**2.12. Morphisms.** Let $V \subseteq \Omega^n$, $W \subseteq \Omega^m$ be algebraic sets. Let us write $\Omega[V] = \Omega[\bar{x}]$, where $\bar{x} = (x_1, \ldots, x_n)$, and each $x_i$ is the image of $X_i$ in $\Omega[V]$.

One can think of an element of $\Omega[V]$ as a function $V \to \Omega$. Indeed, clearly an element $f(\bar{X})$ of $\Omega[\bar{X}]$ defines a function from $\Omega^n$ to $\Omega$. This function restricts to a function $\hat{f}$ defined on $V$. One then has: $\hat{f} = \hat{g}$ if and only if $f - g$ vanishes on $V$ if and only if $f - g \in I(V)$.

A *morphism* $f : V \to W$ is given by a tuple $(f_1(\bar{x}), \ldots, f_m(\bar{x}))$ of elements of $\Omega[V]$, such that for all $\bar{a} \in V$ one has $(f_1(\bar{a}), \ldots, f_m(\bar{a})) \in W$. Note that it suffices to check this for generics of the irreducible components of $V$.

Given $f : V \to W$ as above, we obtain a dual morphism of $\Omega$-algebras $f^* : \Omega[W] \to \Omega[V]$, given by $f^*(g) = g \circ f$, i.e., $f^*(g)(\bar{x}) = g(f_1(\bar{x}), \ldots, f_m(\bar{x})) \in \Omega[V]$.

Assume that $f$, $V$ and $W$ are all defined over $K$, and let $\bar{a}$ be a generic of $V$ over $K$. One has that $f^*$ is injective if and only if $f(V)$ is Zariski dense in $W$, if and only $f(\bar{a})$ is a generic of $W$ over $K$. (The proof is an exercise.) In this case, one says that $f$ is *generically onto*, or *dominant*.

## §3. The theory ACFA. Notation is as in the previous section.

**3.1.** Consider the theory, called ACFA, whose models are the $\mathcal{L}$-structures $K$ satisfying

(1) $K$ is an algebraically closed field;

(2) $\sigma$ is an automorphism of $K$;

(3) If $U$ and $V$ are varieties defined over $K$, with $V \subseteq U \times U^\sigma$, such that the projections of $V$ to $U$ and to $U^\sigma$ are generically onto, then there is a tuple $\bar{a}$ in $K$ such that $(\bar{a}, \sigma(\bar{a})) \in V$.

*Explanation of the notation.*

— $\sigma$ extends to an automorphism of $K[\bar{X}]$ which leaves the elements of $\bar{X}$ fixed. Then $U^\sigma = V(\sigma(I(U)))$, i.e., $U^\sigma \cap K^n = \sigma(U \cap K^n)$.

— The projection maps are induced by $\pi_1 : U \times U^\sigma \to U$ and $\pi_2 : U \times U^\sigma \to U^\sigma$. Our hypothesis simply says that $\pi_1(V)$ is Zariski dense in $U$, and $\pi_2(V)$ is Zariski dense in $U^\sigma$. Equivalently, if whenever $(\bar{a}, \bar{b})$ is a generic of $V$ over $K$, then $\bar{a}$ is a generic of $U$ over $K$, and $\bar{b}$ a generic of $U^\sigma$ over $K$.

*Why (3) is first-order.*

Write $I(U) = (f_1(\bar{X}), \ldots, f_m(\bar{X}))$, and $I(V) = (g_1(\bar{X}, \bar{Y}), \ldots, g_s(\bar{X}, \bar{Y}))$. Choose a tuple $\bar{u}$ and polynomials $F_i(\bar{U}, \bar{X}) \in \mathbb{Z}[\bar{U}, \bar{X}]$, $G_j(\bar{U}, \bar{X}, \bar{Y}) \in \mathbb{Z}[\bar{U}, \bar{X}, \bar{Y}]$ such that $f_i(\bar{X}) = F_i(\bar{u}, \bar{X})$ for $i = 1, \ldots, m$ and $g_j(\bar{X}) = G_j(\bar{u}, \bar{X}, \bar{Y})$ for $j = 1, \ldots, s$.

FACT. The following properties of the tuple $\bar{u}$ are expressible by a first-order formula:

— $F_1(\bar{u}, \bar{X}), \ldots, F_m(\bar{u}, \bar{X})$ generate a prime ideal $I$ in $K[\bar{X}]$;

— $G_1(\bar{u}, \bar{X}, \bar{Y}), \ldots, G_s(\bar{u}, \bar{X}, \bar{Y})$ generate a prime ideal $J$ in $K[\bar{X}, \bar{Y}]$ which intersects $K[\bar{X}]$ in $I$.

The second property tells us that the dual of the projection map gives an inclusion $K[\bar{X}]/I \subseteq K[\bar{X}, \bar{Y}]/J$, which exactly says that $V(J)$ projects generically onto $V(I)$. From the fact, we deduce that each instance of axiom (3) is elementary. (Note that (3) is in fact a scheme of axioms: one for each triple $(n, m, d)$ where $n$ is an upper bound on the number of variables of $u$, $m$ an upper bound on the number of polynomials defining the varieties $U$ and $V$, and $d$ an upper bound on the degree of these polynomials.)

REMARKS. The above fact holds for an arbitrary field $K$, and the formulas expressing the required properties of the tuple $\bar{u}$ do not depend on the field $K$. For a proof, see, e.g., the paper by van den Dries and Schmidt [13]. In this paper there are other very nice (and useful) definability results for ideals in polynomial rings over fields. For example, define uniformly the minimal prime ideals containing an ideal $I$, which when dualized, corresponds to finding the $K$-irreducible components of an algebraic set.

Recall that every formula of the field language is equivalent modulo the theory of algebraically closed fields, to a quantifier-free formula. Hence for instance the property of $\bar{u}$, that $F_1(\bar{u}, \bar{X}), \ldots, F_m(\bar{u}, \bar{X})$ generate a prime ideal in $K^{\mathrm{alg}}[\bar{X}]$, is an elementary property of the tuple $\bar{u}$ in $K$.

**3.2. Theorem.** Every difference field embeds in a model of ACFA. The models of ACFA are exactly the generic difference fields.

PROOF. Let $(K, \sigma)$ be a difference field. Then $\sigma$ lifts to an automorphism of $K^{\mathrm{alg}}$, and so axioms (1) and (2) are no problem. So, let $K$ be an algebraically closed difference field, and let $U$ and $V$ be varieties satisfying the hypotheses of (3). We want to find a difference field $L$ extending $K$ and containing a tuple $\bar{a}$ with $(\bar{a}, \sigma(\bar{a})) \in V$.

Let $(\bar{a}, \bar{b})$ be a generic of $V$ over $K$ (recall, we work in $\Omega$). Then $\bar{a}$ is a generic of $U$ over $K$, and $\bar{b}$ is a generic of $U^{\sigma}$ over $K$. This exactly says that $I(\bar{b}/K) = \sigma(I(\bar{a}/K))$, so that $\sigma$ extends uniquely to a morphism $\tau \colon K(\bar{a}) \to K(\bar{b})$ sending $\bar{a}$ to $\bar{b}$. Let $L = K(\bar{a}, \bar{b})^{\mathrm{alg}}$. By properties of algebraically closed fields, $\tau$ lifts to an automorphism $\rho$ of $L$. Hence $(L, \rho)$ is a difference field extending $(K, \sigma)$ and contains a solution to our equation. $\dashv$

A standard chain argument now shows that every difference field embeds in a model of ACFA (Exercise). We now need to show that the models of ACFA are generic. Let $K$ be a model of ACFA, $f_1(\bar{X}), \ldots, f_m(\bar{X}) \in K[\bar{X}]_\sigma$, and assume that there is a difference field $L$ containing $K$, and a tuple $\bar{a}$ in $L$ such that $f_1(\bar{a}) = \cdots = f_m(\bar{a}) = 0$. We want to show that there is such an $\bar{a}$ in $K$.

Let $\ell \in \mathbb{N}$ be such that $f_1(\bar{X}), \ldots, f_m(\bar{X}) \in K[\bar{X}, \ldots, \sigma^{\ell}(\bar{X})]$. Consider the varieties

— $U$ with generic over $K$ the tuple $\bar{b} = (\bar{a}, \sigma(\bar{a}), \ldots, \sigma^{\ell-1}(\bar{a}))$,

— $V$ with generic over $K$ the tuple $(\bar{b}, \sigma(\bar{b}))$.

Then $\sigma(\bar{b}) = (\sigma(\bar{a}), \ldots, \sigma^{\ell}(\bar{a}))$ is a generic of $U^{\sigma}$. Thus $V \subseteq U \times U^{\sigma}$, and projects generically onto $U$ and onto $U^{\sigma}$. By axiom (3), there is $\bar{c} \in K^{n\ell}$ such that $(\bar{c}, \sigma(\bar{c})) \in V$. Then $\bar{c}$ can be written $(\bar{d}, \sigma(\bar{d}), \ldots, \sigma^{\ell-1}(\bar{d}))$. Since $I(\bar{c}, \sigma(\bar{c})/K)$ contains $I(\bar{b}, \sigma(\bar{b})/K)$, we get that $I(\bar{d}, \sigma(\bar{d}), \ldots, \sigma^{\ell}(\bar{d})/K)$ contains $I(\bar{a}, \sigma(\bar{a}), \ldots, \sigma^{\ell}(\bar{a})/K)$, and therefore that $f_1(\bar{d}) = \cdots = f_m(\bar{d}) = 0$.                                                                 ⊣

**3.3. Corollaries.** (1) ACFA is *model complete*, i.e., if $K_1 \subseteq K_2$ are models of ACFA, then $K_1 \prec K_2$.

(2) Every formula is equivalent, modulo ACFA, to an existential formula.

**3.4. Exercise 2.** Show that if all models of a theory $T$ are existentially closed (among models of $T$), then $T$ is model complete. (Hint of proof: show first that if $A \subseteq B$ and $A$ is existentially closed in $B$, then there is an elementary extension $C$ of $A$ containing $B$. Let now $A \subseteq B$ be models of $T$. Using the first step, construct elementary chains $(A_i)_{i \in \omega}$ and $(B_i)_{i \in \omega}$, with $A_0 = A$, $B_0 = B$, and $A_i \subseteq B_i \subseteq A_{i+1} \subseteq B_{i+1}$. Then $\bigcup_i A_i = \bigcup_i B_i$ is an elementary extension of both $A$ and $B$.)

**3.5. Definition.** Let $E \subseteq K_1, K_2$ be subfields of $\Omega$. One says that $K_1$ and $K_2$ are *algebraically independent over* $E$, or *free over* $E$, if for every $n \in \mathbb{N}$, whenever $a_1, \ldots, a_n \in K_1$ are algebraically independent over $E$, then they remain algebraically independent over $K_2$.

This notion corresponds to independence in the theory of algebraically closed fields, and is a symmetrical notion: $K_1$ is free from $K_2$ over $E$ if and only if $K_2$ is free from $K_1$ over $E$ (see Lang [2]). It is also transitive: if $K_2 \subseteq K_3$, then $K_1$ and $K_3$ are free over $E$ if and only if $K_1$ and $K_2$ are free over $E$, and $K_1 K_2$ and $K_3$ are free over $E$.

**3.6. Definition.** Let $E \subseteq K_1, K_2$ be subfields of $\Omega$. One says that $K_1$ and $K_2$ are *linearly disjoint over* $E$ if for every $n \in \mathbb{N}$, whenever $a_1, \ldots, a_n \in K_1$ and $b_1, \ldots, b_n \in K_2$ are such that $\sum_{i=1}^{n} a_i b_i = 0$ and not all $b_i$s are 0, then there are $c_1, \ldots, c_n \in E$ such that $\sum_{i=1}^{n} a_i c_i = 0$ and not all $c_i$s are 0.

Recall also that the tensor product $K_1 \otimes_E K_2$ is defined as follows: Fix a basis $B_i$ of the $E$-vector space $K_i$, with $1 \in B_i$, $i = 1, 2$. Then, as an $E$-vector space, $K_1 \otimes_E K_2$ has basis

$$\{a \otimes b \mid a \in B_1, \ b \in B_2\}.$$

If $c = \sum_{a \in B_1} c_a a$ and $d = \sum_{b \in B_2} d_b b$ (with the $c_a$ and $d_b$ in $E$; all but finitely many of the $c_a$s and $d_b$s are 0), then we write $c \otimes d$ for the element

$\sum_{a \in B_1, b \in B_2} c_a d_b (a \otimes b)$. Multiplication on the elements of the basis is given by

$$(a \otimes b) \cdot (c \otimes d) = (ac) \otimes (bd),$$

and extended by linearity to the whole space. Note that $K_1$ and $K_2$ embed in $K_1 \otimes_E K_2$, via $a \mapsto a \otimes 1$ and $b \mapsto 1 \otimes b$. If $e \in E$, $a \in K_1$, and $b \in K_2$, then we have $ae \otimes b = e(a \otimes b) = a \otimes eb$.

FACT. $K_1$ and $K_2$ are linearly disjoint over $E$ if and only if $K_1 \otimes_E K_2$ is a domain, equal to $K_1[K_2]$, the subring of $\Omega$ generated by $K_1$ and $K_2$.

For details and proofs, see Lang [2]. The proof of the fact is not difficult. One should note that linear disjointness is a symmetrical notion and implies algebraic independence. It is also transitive. Here are some special cases:

— If $E$ is an algebraically closed field, then $K_1$ and $K_2$ are algebraically independent over $E$ if and only if they are linearly disjoint over $E$.

— If one of $K_1$ or $K_2$ is a Galois extension of $E$, then $K_1$ and $K_2$ are linearly disjoint over $E$ if and only if $K_1 \cap K_2 = E$.

**3.7.** If $E$ is a subset of $K$, one denotes by $\mathcal{L}(E)$ the language obtained by adjoining to $\mathcal{L}$ constant symbols for the elements of $E$. Then $K$ expands naturally to an $\mathcal{L}(E)$-structure, by interpreting the constant symbol corresponding to $e \in E$ by the element $e$ itself. The set of quantifier-free $\mathcal{L}(E)$-sentences true in $K$ is denoted by qfdiag$(E)$, and it describes the isomorphism type of the $\mathcal{L}$-substructure of $K$ generated by $E$. So, a model of qfdiag$(E)$ will be an $\mathcal{L}$-structure containing an isomorphic copy of the $\mathcal{L}$-substructure of $K$ generated by $E$. Elementary equivalence in the language $\mathcal{L}(E)$ is denoted by $\equiv_E$.

THEOREM. Let $K_1$ and $K_2$ be models of ACFA, containing a common algebraically closed difference subfield $E$. Then $K_1 \equiv_E K_2$.

PROOF. To avoid confusion we will denote by $\sigma_i$ the automorphism of $K_i$ that we are considering.

STEP 1. Choose an $E$-isomorphic copy $K_2'$ of the field $K_2$ (by an $E$-isomorphism $\varphi$) which is free from $K_1$ over $E$, and let $\sigma_2' = \varphi \sigma_2 \varphi^{-1}$. Because $E$ is algebraically closed, $K_1$ and $K_2'$ are then linearly disjoint over $E$. Then the difference fields $(K_2, \sigma_2)$ and $(K_2', \sigma_2')$ are $E$-isomorphic via $\varphi$, and therefore elementarily equivalent over $E$. Hence, replacing $K_2$ by $K_2'$, we may assume that $K_1$ and $K_2$ are linearly disjoint over $E$.

STEP 2. We will now show that $\sigma_1$ and $\sigma_2$ have a common extension to $K_1 K_2$ (the subfield of $\Omega$ generated by $K_1$ and $K_2$). We first define $\tau(a \otimes b) = \sigma_1(a) \otimes \sigma_2(b)$ for $a \in K_1$, and $b \in K_2$. Since $\sigma_1$ and $\sigma_2$ agree on $E$, and $K_1$ and $K_2$ are linearly disjoint over $E$, this is well defined. This extends (by linearity) to $K_1 \otimes_E K_2$, which we identify with $K_1[K_2]$, and we then extend $\tau$ to the field of fractions of $K_1[K_2]$, i.e., to $K_1 K_2$.

STEP 3. The difference field $(K_1 K_2, \tau)$ embeds in a model $L$ of ACFA. Because ACFA is model complete (Corollary 3.3(1)) we then have $K_i \prec L$ for $i = 1, 2$, which implies that $K_1$ and $K_2$ satisfy the same $\mathcal{L}(E)$-sentences, and shows the result.                                                                 ⊣

**3.8. Corollary.** Let $E$ be an algebraically closed difference field. Then ACFA $\cup$ qfdiag$(E)$ is complete.

PROOF. If $K_1$ and $K_2$ are models of ACFA $\cup$ qfdiag$(E)$, then $K_1$ and $K_2$ contain difference subfields $E_1$ and $E_2$, respectively, which are isomorphic to $E$. Moving $K_2$ by an isomorphism, we may assume that $E_1 = E_2$. The result then follows by Theorem 3.7.                                              ⊣

**3.9. Corollary.** The completions of ACFA are obtained by adjoining to ACFA a description of the isomorphism type of the difference field consisting of elements algebraic over the prime field.

PROOF. If $T = \mathrm{Th}(K)$ is a complete theory containing ACFA, then $T$ will specify the characteristic, and therefore the isomorphism type of the prime field $k$. Note that the elements of $k$ are in fact (interpretations of) terms of the language $\mathcal{L}$. Let $L$ be a finite Galois extension of $k$ of degree $n$ over $k$, $\alpha$ a generator of $L$ over $k$, and $p(X) \in k[X]$ its minimal (monic) polynomial. Since $L$ is Galois over $k$, and $k$ is fixed by $\sigma$, $\sigma(L)$ will contain all the roots of $p(X)$, and therefore will equal $L$. Hence, $\sigma(\alpha) = \sum_{i=0}^{n-1} a_i \alpha^i$ for some elements $a_i \in k$. The sentence

$$\exists x \; p(x) = 0 \wedge \sigma(x) = \sum_{i=0}^{n-1} a_i x^i$$

will therefore belong to $T$. The set of all such sentences will describe the isomorphism type of the difference subfield $k^{\mathrm{alg}}$ of $K$.

The converse follows from Theorem 3.7.                                        ⊣

**3.10. Notation.** If $A$ is a subset of an algebraically closed difference field $K$, we denote by $\mathrm{cl}_\sigma(A)$, the difference field generated by $A$, and by $\mathrm{acl}_\sigma(A)$ the smallest algebraically closed difference field containing $A$. Note that $\mathrm{cl}_\sigma(A)$ is the field generated by the sets $\sigma^i(A), i \in \mathbb{Z}$, and that $\mathrm{acl}_\sigma(A)$ is simply the algebraic (field-theoretic) closure of $\mathrm{cl}_\sigma(A)$.

Recall that the model-theoretic *definable and algebraic closure* are defined as follows: let $M$ be a model of a theory, and $A \subset M$. The *definable closure* of $A$ in $M$, denoted $\mathrm{dcl}(A)$, is the set of elements $a \in M$ such that there is some formula $\varphi(x) \in \mathcal{L}(A)$ which is satisfied by $a$ in $M$ and by no other element of $M$. We will then say that the formula $\varphi(x)$ defines $a$. So, $\mathrm{dcl}(A)$ will in particular contain all elements of the substructure of $M$ generated by $A$. The *algebraic closure* of $A$ in $M$, denoted $\mathrm{acl}(A)$, is the set of elements $a \in M$ which satisfy some $\mathcal{L}(A)$-formula which is satisfied by only finitely many elements of $M$.

Clearly $dcl(A) \subseteq acl(A)$, and $A \subseteq B$ implies $dcl(A) \subseteq dcl(B)$ and $acl(A) \subseteq acl(B)$. Moreover, one can show that $acl(acl(A)) = acl(A)$ (and, of course, $dcl(dcl(A)) = dcl(A)$). In the particular case of difference fields, clearly $cl(A) \supseteq cl_\sigma(A)$, and $acl(A) \supseteq acl_\sigma(A)$.

**3.11. Corollary.** Let $(K_1, \sigma_1)$ and $(K_2, \sigma_2)$ be models of ACFA containing a common difference subfield $(E, \sigma)$. Then

$$K_1 \equiv_E K_2 \iff (E^{alg}, \sigma_1|_{E^{alg}}) \simeq_E (E^{alg}, \sigma_2|_{E^{alg}}).$$

PROOF. The left to right implication is clear, as elements of $E^{alg}$ are algebraic over $E$. For the converse, let $\varphi\colon E^{alg} \to E^{alg}$ be an $E$-isomorphism such that $\varphi\sigma_1 = \sigma_2\varphi$. Extend $\varphi$ to an automorphism $\psi$ of $\Omega$, and let $(K_1', \sigma_1') = (\psi(K_1), \psi\sigma_1\psi^{-1})$ be the difference field image of $K_1$ by $\psi$. Then $K_1 \equiv_E K_1'$, and $\sigma_1'$ and $\sigma_2$ agree on $E^{alg}$. So we may apply Theorem 3.7.     ⊣

**3.12. Corollary.** Let $\varphi(\bar{x})$ be a formula. Then, modulo ACFA, $\varphi(\bar{x})$ is equivalent to a disjunction of formulas of the form $\exists y\ \psi(\bar{x}, y)$, where $\psi(\bar{x}, y)$ is quantifier free, and for every difference field $K$ and $(\bar{a}, b)$ in $K$ satisfying $\psi$, we have that $b$ is algebraic over $(\bar{a}, \sigma(\bar{a}), \ldots)$.

**3.13. Exercise 3.** Give a proof of Corollary 3.12. (Hint: First note that if $E$ is a separably closed difference field, then $qfdiag(E) \vdash qfdiag(E^{alg})$ modulo the theory of difference fields. Then, for every model $K$ of ACFA, and tuple $\bar{a}$ satisfying $\varphi$, find a formula $\psi_a$ implying $\varphi$ and of the required form. Use compactness to conclude.)

**3.14. Definition of types and saturated models.** Let $M$ be an $\mathcal{L}$-structure, $A$ a subset of $M$, and $\bar{a}$ an $n$-tuple of elements of $M$. The *type of $\bar{a}$ over $A$ in $M$*, denoted $tp(\bar{a}/A)$ or sometimes $tp_M(\bar{a}/A)$, is the set of $\mathcal{L}(A)$-formulas $\varphi(\bar{x})$ satisfied by $\bar{a}$ in $M$, where $\bar{x}$ is a fixed $n$-tuple of variables. A *partial $n$-type over $A$* is a set $\Gamma(\bar{x})$ of $\mathcal{L}(A)$-formulas in the variables $\bar{x}$ such that every finite conjunction of elements of $\Gamma(\bar{x})$ is satisfiable by some $n$-tuple of $M$. By compactness, a partial $n$-type over $A$ will be realized in some elementary extension $M^*$ of $M$, i.e., $M^*$ will contain some tuple $\bar{a}$ which satisfies all formulas of $\Gamma(\bar{x})$.

If $\kappa$ is an infinite cardinal, an $\mathcal{L}$-structure $M$ is *$\kappa$-saturated* if for every $A \subseteq M$ with $|A| < \kappa$, every (partial) type over $A$ is realized in $M$. The structure $M$ is saturated if and only if it is $|M|$-saturated. Saturated models are quite useful, as they realize many types and have many automorphisms: if $M$ is saturated and $f\colon A \to B$ is an elementary map between two subsets of $M$ of size $< |M|$, then $f$ extends to an automorphism of $M$.

**3.15. Corollary.** Let $E$ be a difference subfield of a model $K$ of ACFA, and let $\bar{a}$ and $\bar{b}$ be tuples in $K$ of the same length. Then $tp(\bar{a}/E) = tp(\bar{b}/E)$ if and only if there is an $E$-isomorphism (of difference fields) $acl_\sigma(E\bar{a}) \to acl_\sigma(E\bar{b})$ sending $\bar{a}$ to $\bar{b}$.

PROOF. Extend the $E$-isomorphism $\mathrm{acl}_\sigma(E\bar{a}) \rightarrow \mathrm{acl}_\sigma(E\bar{b})$ to an $E$-isomorphism $\varphi : K \rightarrow K_1$ (for some difference field $K_1$). Then $\mathrm{tp}_K(\bar{a}/E) = \mathrm{tp}_{K_1}(\bar{b}/E)$ (since $\varphi(\bar{a}) = \bar{b}$). By Theorem 3.7, $K_1 \equiv_{\mathrm{acl}_\sigma(E\bar{b})} K$, which implies that $\mathrm{tp}(\bar{a}/E) = \mathrm{tp}(\bar{b}/E)$.                          ⊣

**3.16. Corollary.** Let $K$ be a model of ACFA, and $A$ a subset of $K$. Then $\mathrm{acl}_\sigma(A)$ equals the model-theoretic algebraic closure of $A$, $\mathrm{acl}(A)$.

PROOF. Clearly $\mathrm{acl}_\sigma(A) \subseteq \mathrm{acl}(A)$. Choose an $\mathrm{acl}_\sigma(A)$-isomorphic copy $K_1$ of $K$, which is linearly disjoint from $K$ over $\mathrm{acl}_\sigma(A)$. As in Theorem 3.7, $KK_1$ embeds in a model $L$ of ACFA. By Corollary 3.15, if $a \in K \setminus \mathrm{acl}_\sigma(A)$, then there is $b \in K_1 \setminus \mathrm{acl}_\sigma(A)$ realizing the same type over $\mathrm{acl}_\sigma(A)$ as $a$. Hence no type realized in $K \setminus \mathrm{acl}_\sigma(A)$ is algebraic.                          ⊣

**3.17. Saturated models of ACFA.** Let $K$ be a saturated model of ACFA. Then $K$ has the following property: if $E$ is an algebraically closed difference subfield of $K$, and $F$ is a difference field extending $E$, and with $|F| < |K|$, then there is an $E$-embedding of $F$ into $E$.

Thus saturated models of ACFA, of large enough cardinality, play the role of universal models of algebraic geometry.

## §4. $\sigma$-closed sets, independence, and SU-rank.
We keep the notation and conventions introduced before. We work in a (sufficiently saturated) model $K$ of ACFA.

**4.1. $\sigma$-closed sets and the topology they generate.** In analogy with the Zariski topology, we define the $\sigma$-topology on $K^n$. Given $B \subset K[\bar{X}]_\sigma$, $\bar{X} = (X_1, \ldots, X_n)$, we set

$$V(B) = \{\bar{a} \in K^n \mid f(\bar{a}) = 0 \text{ for all } f(\bar{X}) \in B\}.$$

Dually, if $S \subseteq K^n$, we define

$$I_\sigma(S) = \{f(\bar{X}) \in K[\bar{X}]_\sigma \mid f(\bar{a}) = 0 \text{ for all } \bar{a} \in S\}.$$

If $E$ is a difference subfield of $K$ and $\bar{a} \in K^n$, we define

$$I_\sigma(\bar{a}/E) = \{f(\bar{X}) \in E[\bar{X}]_\sigma \mid f(\bar{a}) = 0\}.$$

We call the sets of the form $V(B)$ the $\sigma$-closed subsets of $K^n$. One checks easily that $V(I_\sigma(V(B))) = V(B)$. Some of the following results are theorems (see Cohn [1]):

FACTS AND REMARKS. Note that $I_\sigma(S)$ is an ideal $I$ of $K[\bar{X}]$ with the following properties:

(i) $f \in I \iff \sigma(f) \in I$.

(ii) If $f^m \sigma(f)^n \in I$, then $f \in I$.

Ideals satisfying (i) are called *reflexive $\sigma$-ideals* (and $\sigma$ will then induce an injective endomorphism on the quotient of $K[\bar{X}]$ by such an ideal). Ideals satisfying in addition condition (ii) are called *perfect $\sigma$-ideals*. Prime ideals satisfying (i) and (ii) are called *prime $\sigma$-ideals*. Perfect $\sigma$-ideals are the analogues of radical ideals and are intersections of prime $\sigma$-ideals.

**4.2. Fact.** Even though $K[\bar{X}]_\sigma$ does not satisfy the a.c.c. on $\sigma$-ideals, it satisfies it on perfect $\sigma$-ideals, and this implies that the $\sigma$-topology on $K^n$ is Noetherian.

**4.3. Exercise 4.** Show that if $I$ is a prime $\sigma$-ideal then $I_\sigma(V(I)) = I$. (Warning: your proof should use the fact that $K$ is a model of ACFA. You may use the fact that a field with a distinguished endomorphism embeds in an inversive difference field.)

**4.4. Definition.** Let $E$ be a difference subfield of $K$ and $a \in K$. We say that $a$ is *transformally transcendental* over $E$ if $I_\sigma(a/E) = 0$. Otherwise, we say that $a$ is *transformally algebraic* over $E$. A tuple is *transformally algebraic* over $E$ if all its elements are.

If $a$ is transformally transcendental over $E$, then the elements $\sigma(a), i \in \mathbb{N}$, are algebraically independent over $E$. Hence, applying $\sigma^{-1}$, so are the elements $\sigma(a), i \in \mathbb{Z}$. Thus, the difference field generated by $a$ over $E$ is isomorphic to $E(X^{\sigma^i} \mid i \in \mathbb{Z})$ (with the obvious action of $\sigma$).

Similarly, one says that $n$-tuple $\bar{a}$ is *transformally independent* over $E$, if $I_\sigma(\bar{a}/E) = (0)$. There are notions of *transformal transcendental bases, transformal transcendental degree* of an extension, etc.

Assume now that $a$ is transformally algebraic over $E$, and let $m$ be least such that some $f(X) = F(X, X^\sigma, \ldots, X^{\sigma^m}) \in I_\sigma(a/E)$. Choose such an $f(X)$ of lowest degree when viewed as a polynomial in $X^{\sigma^m}$. Then $F(a, \ldots, \sigma^{m-1}(a), Y)$ is irreducible over $E(a, \ldots, \sigma^{m-1}(a))$ because $I_\sigma(a/E) \cap E[X, \ldots, X^{\sigma^m}]$ is prime, and is the minimal polynomial of $\sigma^m(a)$ over $E(a, \ldots, \sigma^{m-1}(a))$.

From $F(a, \ldots, \sigma^m(a)) = 0$, we deduce that $\sigma(F)(\sigma(a), \ldots, \sigma^{m+1}(a)) = 0$, so that the minimal polynomial of $\sigma^{m+1}(a)$ over $E(a, \ldots, \sigma^m(a))$ divides $\sigma(F)(\sigma(a), \ldots, \sigma^m(a), Y)$, and therefore has degree bounded above by the degree of $F(a, \ldots, \sigma^{m-1}(a), Y)$. It follows that $I_\sigma(a/E)$, as a $\sigma$-ideal, is finitely generated, since from some point on, the degree of the minimal polynomial of $\sigma^n(a)$ over $E(a, \ldots, \sigma^{n-1}(a))$ must stabilize.

Note that we also have that
$$\mathrm{acl}_\sigma(Ea) = E(a, \ldots, \sigma^{m-1}(a))^{\mathrm{alg}}.$$

Similarly, if $\bar{a}$ is transformally algebraic over $E$, then there is an $n$ such that $\mathrm{acl}_\sigma(E\bar{a}) = E(\bar{a}, \ldots, \sigma^n(\bar{a}))^{\mathrm{alg}}$.

**4.5. Exercise 5.**

(1) Use the preceding remarks to show that every prime $\sigma$-ideal of $E[\bar{X}]_\sigma$, is finitely generated (as a $\sigma$-ideal).

(2) (harder) Let $I$ be a perfect $\sigma$-ideal of $K[\bar{X}]$, $\bar{X} = (X_1, \ldots, X_n)$, and assume that $I \cap K[X_i]_\sigma \neq 0$ for all $i = 1, \ldots, n$. Show that any descending sequence of $\sigma$-closed subsets $F_j$ of $V(I)$ is finite (Hint: Look at the

Zariski closures of $\{(\bar{a}, \sigma(\bar{a}), \ldots, \sigma^m(\bar{a})) \mid \bar{a} \in F_j\}$ for $m$ large enough; use induction on the dimension of the algebraic sets considered.)

(3) Using (2), show that the $\sigma$-topology is Noetherian.

**4.6. Definition.** Let $A$, $B$, $C$ be subsets of a model $K$ of ACFA. We say that $A$ and $B$ are *independent over* $C$ if the fields $\mathrm{acl}_\sigma(CA)$ and $\mathrm{acl}_\sigma(CB)$ are free (or equivalently, linearly disjoint) over $\mathrm{acl}_\sigma(C)$. We denote it by $A \downarrow_C B$.

REMARKS. Independence is clearly a symmetrical notion, and is transitive, i.e., let $B' \subset B$. Then

$$A \downarrow_C B \iff A \downarrow_C B' \text{ and } A \downarrow_{C \cup B'} B.$$

Note also that by definition

$$A \downarrow_C B \iff \mathrm{acl}_\sigma(C, A) \downarrow_{\mathrm{acl}_\sigma(C)} \mathrm{acl}_\sigma(C, B)$$

and that

$$\mathrm{acl}_\sigma(C, A) = (\mathrm{acl}_\sigma(C)\mathrm{acl}_\sigma(A))^{\mathrm{alg}}.$$

Assume for simplicity, that $C \subseteq B$ are algebraically closed difference fields. Then $A$ and $B$ are independent over $C$ if and only if, for every tuple $\bar{a} \in A$, the ideal $I_\sigma(\bar{a}/B)$ is generated by its intersection with $C[\bar{X}]_\sigma$.

Moreover, independence satisfies the extension property: given $A$, $B$, and $C$, there is $A'$ realizing $\mathrm{tp}(A/C)$ in some elementary extension of $K$ such that $A'$ and $B$ are independent over $C$. Indeed, without loss of generality, we may assume that $C$, $A$ and $B$ are algebraically closed difference fields, with $C \subset A \cap B$. Let $A'$ be a $C$-isomorphic copy of $A$ which is free from $B$ over $C$. Then $A'$ is linearly disjoint from $B$ over $C$, reasoning as in Step 2 of Theorem 3.7, we get that there is an elementary extension $L$ of $K$ which contains the difference fields $A'$ and $B$. Then $A'$ and $B$ are independent over $C$, and $\mathrm{tp}_L(A'/C) = \mathrm{tp}_L(A/C)$.

**4.7. Exercise 6.** Let $K$ be a model of ACFA, and $A$, $B$, $C$, $D$ algebraically closed difference subfields of $K$. Give a proof, or convince yourself, of the following facts:

(1) (Symmetry) If $A \downarrow_C B$ then $B \downarrow_C A$.

(2) (Transitivity) If $C \subseteq A \cap B$ and $B \subseteq D$,

$$A \downarrow_C D \iff A \downarrow_C B \text{ and } A \downarrow_B D.$$

(3) $A \downarrow_C B$ if and only if for every finite tuple $\bar{b}$ from $B$, $A \downarrow_C \bar{b}$.

(4) (Extension) In some elementary extension of $K$ there is $A'$ realizing $\mathrm{tp}(A/C)$ with $A' \downarrow_C B$. (This was already explained above.)

(5) There is a finite subset $E$ of $C$ such that $A \downarrow_E C$.

**4.8. The independence theorem.** Let $K$ be a model of ACFA (sufficiently saturated), let $E = \mathrm{acl}_\sigma(E) \subseteq K$, and $\bar{a}$, $\bar{b}$, $\bar{c}_1$ and $\bar{c}_2$ tuples in $K$ such that

(i) $\mathrm{tp}(\bar{c}_1/E) = \mathrm{tp}(\bar{c}_2/E)$,

(ii) $\bar{a}$ and $\bar{c}_1$ are independent over $E$, $\bar{a}$ and $\bar{b}$ are independent over $E$ and $\bar{b}$ and $\bar{c}_2$ are independent over $E$.

Then there is $\bar{c}$ realizing $\mathrm{tp}(\bar{c}_1/E \cup \bar{a}) \cup \mathrm{tp}(\bar{c}_2/E \cup \bar{b})$, independent from $(\bar{a}, \bar{b})$ over $E$.

PROOF. Let $\bar{c}$ realize $\mathrm{tp}(\bar{c}_1/E)$, independent from $(\bar{a}, \bar{b})$ over $E$. Let $A = \mathrm{acl}_\sigma(E\bar{a})$, $B = \mathrm{acl}_\sigma(E\bar{b})$, $C = \mathrm{acl}_\sigma(E\bar{c})$, and fix $E$-isomorphisms (of difference fields) $\varphi_1: \mathrm{acl}_\sigma(E\bar{c}_1) \to C$ and $\varphi_2: \mathrm{acl}_\sigma(E\bar{c}_2) \to C$, with $\varphi_i(\bar{c}_i) = \bar{c}$.

Let $\sigma_0$ be the restriction of $\sigma$ to $(AB)^{\mathrm{alg}}C$. Since $A$ is linearly disjoint from $\mathrm{acl}_\sigma(E\bar{c}_1)$ and from $C$ over $E$, we may extend $\varphi_1$ to a *field*-isomorphism $\psi_1: \mathrm{acl}_\sigma(A\bar{c}_1) \to (AC)^{\mathrm{alg}}$ which is the identity on $A$. Then $\sigma_1 = \psi_1\sigma\psi_1^{-1}$ is an automorphism of $(AC)^{\mathrm{alg}}$ which agrees with $\sigma$ on $A$ and on $C$. Indeed, $\sigma_1$ agrees with $\sigma$ on $A$ because $\psi_1$ is the identity on $A$, and on $C$ because $\psi_1$ extends the difference field isomorphism $\varphi_1: \mathrm{acl}_\sigma(E\bar{c}_1) \to C$. Note also that by definition of $\sigma_1$, the isomorphism $\psi_1$ is an isomorphism between the difference field $(\mathrm{acl}_\sigma(A\bar{c}_1), \sigma)$ and the difference field $((AC)^{\mathrm{alg}}, \sigma_1)$.

Similarly, we may extend $\varphi_2$ to a field-isomorphism $\psi_2: \mathrm{acl}_\sigma(B\bar{c}_2) \to (BC)^{\mathrm{alg}}$ which is the identity on $B$. The automorphism $\sigma_2 = \psi_2\sigma\psi_2^{-1}$ agrees with $\sigma$ on $B$ and $C$. $\dashv$

Assume that there is an automorphism $\tau$ of $L = (AB)^{\mathrm{alg}}(AC)^{\mathrm{alg}}(BC)^{\mathrm{alg}}$ which extends $\sigma_0$, $\sigma_1$, and $\sigma_2$. Then we can find some model $M$ of ACFA containing $(L, \tau)$. As $\tau$ extends $\sigma_0$, we then have that $\mathrm{tp}_M(AB/E) = \mathrm{tp}_K(AB/E)$ (by Theorem 3.7). By Corollary 3.15, we also have

$$\mathrm{tp}_M(\bar{c}/A) = \mathrm{tp}_K(\bar{c}_1/A) \quad \text{and} \quad \mathrm{tp}_M(\bar{c}/B) = \mathrm{tp}_K(\bar{c}_2/B)$$

because $\tau$ extends $\sigma_1$ and $\sigma_2$, and the $\psi_i$ are difference field isomorphisms fixing $A$ and $B$, respectively. Clearly, $\bar{c}$ is independent from $(A, B)$ over $E$, and this will have finished the proof.

It remains to show that there is such an automorphism $\tau$ of $L$. To do that, it suffices to show that $\sigma_0$ and $\sigma_1$ have a common (and necessarily unique) extension $\tau_1$ to $(AB)^{\mathrm{alg}}(AC)^{\mathrm{alg}}$ and that $\tau_1$ and $\sigma_2$ have a common extension to $L$.

To show that $\sigma_0$ and $\sigma_1$ have a common extension $\tau_1$ to $(AB)^{\mathrm{alg}}(AC)^{\mathrm{alg}}$, it is enough to show that their domains are linearly disjoint over their intersection, and that $\sigma_0$ and $\sigma_1$ agree on this intersection. Similarly for $\tau_1$ and $\sigma_2$.

The domain of $\sigma_0$ is $(AB)^{\mathrm{alg}}C$, the domain of $\sigma_1$ is $(AC)^{\mathrm{alg}}$ which is a Galois extension of $AC$. By definition, $\sigma_0$ is the restriction of $\sigma$ ($\in \mathrm{Aut}(K)$), and we know that $\sigma_1$ and $\sigma$ agree on $AC$. It follows that it suffices to show that

$$(AB)^{\mathrm{alg}}C \cap (AC)^{\mathrm{alg}} = AC. \tag{1}$$

(Here we are using the fact that $(AC)^{\text{alg}}$ is a Galois extension of $AC$ to reduce the linear disjointness over $AC$ to intersecting in $AC$.) Similarly, to show that $\tau_1$ and $\sigma_2$ have a common extension to $L$, it will be enough to show that

$$(AB)^{\text{alg}}(AC)^{\text{alg}} \cap (BC)^{\text{alg}} = BC. \tag{2}$$

Unfortunately the proof of either of these equations uses tools slightly beyond the scope of this course, since I had chosen not to assume anything known in stability theory. For sake of completeness I will give the proof using stability results. The reader unfamiliar with stable theories may skip this part and admit the equations (1) and (2), and therefore the result. Let us prove (2) first. Algebraists would prove it via specializations; the ideas are essentially the same (but not the way it is stated).

Let $\alpha \in (BC)^{\text{alg}} \cap (AB)^{\text{alg}}(AC)^{\text{alg}}$, and write $\alpha = \sum_{i=1}^{n} \beta_i \gamma_i$, where $\beta_i \in (AB)^{\text{alg}}, \gamma_i \in (AC)^{\text{alg}}$. Let $\bar{a}'$, $\bar{b}'$, and $\bar{c}'$ be tuples of elements from $A$, $B$, and $C$, respectively, and $f_i(\bar{X}, \bar{Y}, U), g_i(\bar{X}, \bar{Z}, V)$ be polynomials over $E$ such that $f_i(\bar{a}', \bar{b}', U)$ is the minimal polynomial of $\beta_i$ over $AB$, and $g_i(\bar{a}', \bar{c}', V)$ is the minimal polynomial of $\gamma_i$ over $AC$ for $i = 1, \ldots, n$. Then

$$K \models \exists u_1, \ldots, u_n, v_1, \ldots, v_n$$
$$\bigwedge_i \left( f_i(\bar{a}', \bar{b}', u_i) = 0 \wedge g_i(\bar{a}', \bar{c}', v_i) = 0 \right) \wedge \alpha = \sum_i u_i v_i.$$

Note that this is a formula of the field language satisfied by $(\bar{a}', \bar{b}', \bar{c}', \alpha)$. By assumption, $\bar{a}'$ is independent from $(BC)^{\text{alg}}$ over $E$, in the sense of the theory of algebraically closed fields. Hence every formula represented in $\text{tp}_{\text{ACF}}(\bar{b}', \bar{c}', \alpha/A)$ is already represented in $\text{tp}_{\text{ACF}}(\bar{b}', \bar{c}', \alpha/E)$ (here $\text{tp}_{\text{ACF}}$ denotes the type in the theory of algebraically closed fields). This precisely means that there is a tuple $\bar{e}$ in $E$ such that

$$K \models \exists u_1, \ldots, u_n, v_1, \ldots, v_n$$
$$\bigwedge_i \left( f_i(\bar{e}, \bar{b}', u_i) = 0 \wedge g_i(\bar{e}, \bar{c}', v_i) = 0 \right) \wedge \alpha = \sum_i u_i v_i.$$

If $\beta_1', \ldots, \beta_n', \gamma_1', \ldots, \gamma_n'$ satisfy $\bigwedge_i (f_i(\bar{e}, \bar{b}', \beta_i') = 0 \wedge g_i(\bar{e}, \bar{c}', \gamma_i') = 0) \wedge \alpha = \sum_i \beta_i' \gamma_i'$, then $\beta_i' \in B$ and $\gamma_i' \in C$, which shows that $\alpha \in BC$ and proves (2). Permuting $A$, $B$, $C$, we obtain that $(AC)^{\text{alg}} \cap (AB)^{\text{alg}}(BC)^{\text{alg}} = AC$ from which we get (1).

**4.9. Corollary.** $\square\square\square$  All completions of ACFA are supersimple, and independence corresponds to nonforking.

PROOF. See the paper by Kim and Pillay [15]. They show that if you have an independence notion which is

(i) symmetric,
(ii) transitive,

(iii) has the extension property,
(iv) is such that, given a finite tuple $\bar{a}$ and a set $A$, there is $A_0 \subseteq \mathrm{acl}^{\mathrm{eq}}(A)$ such that $\bar{a}$ and $A$ are independent over $A_0$, and $|A_0| \leq |\mathcal{L}| + \aleph_0$,
(v) satisfies the independence theorem.

Then this notion of independence coincides with nonforking and the theory is simple. If in (iv) the set $A_0$ can always be taken finite, then the theory is supersimple.

In our case, given the complete theory of a model $K$ of ACFA, items (i)–(iii) are immediate because similar statements hold for nonforking in algebraically closed fields, (v) is Theorem 4.8. From our definition of independence and the descending chain condition on $\sigma$-closed sets (Fact 4.2), it follows that in (iv) we can always take $A_0$ to be finite.  □□□

**4.10. Definitions of forking and of the SU-rank.** The SU-rank is a rank on types, based on nonindependence in the same way the U-rank is.

Let $K$ be a model of ACFA, sufficiently saturated, $E \subseteq F$ algebraically closed subsets of $K$, and $\bar{a}$ a tuple of elements, $p = \mathrm{tp}(\bar{a}/E)$, $q = \mathrm{tp}(\bar{a}/F)$. We say that $q$ *forks* over $E$, or that $q$ *is a forking extension of $p$*, if and only if $\bar{a} \not\downarrow_E F$. Otherwise, we say that $q$ *does not fork over $E$*, or that $q$ *is a nonforking extension of $p$ to $F$*.

We define $\mathrm{SU}(p) = \mathrm{SU}(\bar{a}/E) \geq \alpha$ by induction on the ordinal $\alpha$:

— $\mathrm{SU}(p) \geq 0$;

— $\mathrm{SU}(p) \geq \alpha + 1$ if and only if $p$ has a forking extension $q$ such that $\mathrm{SU}(q) \geq \alpha$, if and only if there is $B = \mathrm{acl}_\sigma(B)$ containing $E$ such that $\mathrm{acl}_\sigma(E\bar{a})$ and $B$ are not linearly disjoint over $\mathrm{acl}_\sigma(E)$, and $\mathrm{SU}(\bar{a}/B) \geq \alpha$;

— if $\alpha$ is a limit ordinal, then $\mathrm{SU}(p) \geq \alpha$ if and only if $\mathrm{SU}(p) \geq \beta$ for all $\beta < \alpha$.

We then define $\mathrm{SU}(p)$ to be the least $\alpha$ such that $\mathrm{SU}(p) \not\geq \alpha + 1$ if it exists, and $\infty$ otherwise.

**4.11. Exercise 7.** Let $\bar{a}, E \subseteq F, K$ be as above. Give a proof, or convince yourself of the following facts:

(1) $\mathrm{SU}(\bar{a}/F) \leq \mathrm{SU}(\bar{a}/E)$. (Hint: show by induction on $\alpha$ that $\mathrm{SU}(\bar{a}/F) \geq \alpha$ implies $\mathrm{SU}(\bar{a}/E) \geq \alpha$.)

(2) if $\bar{a} \downarrow_E F$, then $\mathrm{SU}(\bar{a}/E) = \mathrm{SU}(\bar{a}/F)$. (Again, use induction on $\alpha$. The independence theorem intervenes in the proof.)

(3) Deduce from the remark in Subsection 4.6 and the Noetherianity of the $\sigma$-topology that $\mathrm{SU}(\bar{a}/E) < \infty$.

**4.12. Remark.** As explained in the exercise, Fact 4.2 yields that the SU-rank of a type (of a finite tuple) exists. It is, however, much stronger: you could imagine that there could exist an infinite descending sequence of $\sigma$-closed sets all defined over $\mathrm{acl}_\sigma(\emptyset)$.

**4.13. Natural sum on ordinals.** Every ordinal can be written uniquely as

$$\alpha = \omega^{\alpha_1} a_1 + \cdots + \omega^{\alpha_n} a_n,$$

where $\alpha_1 > \cdots > \alpha_n \geq 0$ are ordinals, and $a_1, \ldots, a_n$ are positive integers. If $\beta = \omega^{\beta_1} b_1 + \cdots + \omega^{\beta_m} b_m$, then we define $\alpha \oplus \beta$ as follows. First, relaxing the condition on $a_1, \ldots, a_n, b_1, \ldots, b_m$ to be positive and allowing them to be 0, we may assume that $m = n$ and $\alpha_i = \beta_i$ for $i = 1, \ldots, n$. Then one sets

$$\alpha \oplus \beta = \omega^{\alpha_1}(a_1 + b_1) + \cdots + \omega^{\alpha_n}(a_n + b_n).$$

One verifies that $\oplus$ is commutative and transitive.

While $\oplus$ coincides with the usual ordinal addition on finite ordinals, it definitely does not on infinite ones. For instance $1 \oplus \omega = \omega \oplus 1 = \omega + 1$, but $1 + \omega = \omega$.

**4.14. Properties of the SU-rank.** One can show that the SU-rank satisfies the so-called Lascar inequality: given another tuple $\bar{b}$,

$$\mathrm{SU}(\bar{a}/E\bar{b}) + \mathrm{SU}(\bar{b}/E) \leq \mathrm{SU}(\bar{a}, \bar{b}/E) \leq \mathrm{SU}(\bar{a}/A\bar{b}) \oplus \mathrm{SU}(\bar{b}/A).$$

This is shown by induction. For the first inequality, one shows that $\mathrm{SU}(\bar{b}/E) \geq \alpha$ implies $\mathrm{SU}(\bar{a}/E\bar{b}) + \alpha \geq \mathrm{SU}(\bar{a}, \bar{b}/E)$. For the second, that $\mathrm{SU}(\bar{a}, \bar{b}/E) \geq \alpha$ implies $\mathrm{SU}(\bar{a}/E\bar{b}) \oplus \mathrm{SU}(\bar{a}/E) \geq \alpha$. The proof is not difficult, but one can also consult, e.g., Wagner's *Simple theories* [19].

**4.15. SU-rank of definable sets or of formulas.** Let $S \subseteq K^n$ be a definable set defined by a formula $\varphi(\bar{x})$ over some $E$, and assume that $K$ is sufficiently saturated (otherwise our definition will not make sense). We define $\mathrm{SU}(\varphi) = \mathrm{SU}(S) = \sup\{\mathrm{SU}(\bar{a}/A) \mid \bar{a} \in S\}$. One can show that this sup is attained, i.e., that there is some $\bar{a}$ satisfying $\varphi$ and such that $\mathrm{SU}(\bar{a}/A) = \mathrm{SU}(\varphi)$.

**4.16. Remarks.** Note the following special cases:

$$\mathrm{SU}(\bar{a}/E) = 0 \iff \bar{a} \in \mathrm{acl}_\sigma(E);$$
$$\mathrm{SU}(\bar{a}/E) = 1 \iff \bar{a} \notin \mathrm{acl}_\sigma(E) \quad \text{and for all } F \supset E,$$
$$\text{either } \bar{a} \text{ is independent from } F \text{ over } E, \text{ or } \bar{a} \in \mathrm{acl}_\sigma(F).$$

To simplify, let us assume that $E$ is an algebraically closed difference field (since $\mathrm{SU}(\bar{a}/\mathrm{acl}_\sigma(E)) = \mathrm{SU}(\bar{a}/E)$). Let us also assume for the moment that the tuple $\bar{a}$ is transformally algebraic over $E$, that is, that $\mathrm{tr.deg}(\mathrm{cl}_\sigma(E\bar{a})/E) = n$ is *finite*. Then, $\mathrm{tp}(\bar{a}/F)$ forks over $E$ if and only if

$$\mathrm{tr.deg}(\mathrm{cl}_\sigma(F\bar{a})/F) < \mathrm{tr.deg}(\mathrm{cl}_\sigma(E\bar{a})/E).$$

This implies that $\mathrm{SU}(\bar{a}/E) \leq n$. Equality, however, does not always hold. For instance, one can show that any nonrealized type containing the formula $\sigma^2(x) = x^2$, or the formula $\sigma^2(x) = x^2 + 1$, has SU-rank 1.

**4.17. The type of SU-rank $\omega$.** Let $E = \mathrm{acl}_\sigma(E)$, and consider an element $a \in K$ which is transformally transcendental over $E$, that is, the elements $\sigma^i(a), i \in \mathbb{Z}$, are algebraically independent over $E$. Let $b_0 = a$, and $b_n = \sigma(b_{n-1}) - b_{n-1}$ for $n \geq 1$. Then, letting $L_n = \mathrm{cl}_\sigma(Eb_n)$, we get that each $L_n$ contains $L_{n+1}$, and has transcendence degree 1 over $L_{n+1}$. Hence, $\mathrm{SU}(b_n/Eb_{n+1}) = 1$ (since it is not 0 and is at most 1). Hence, $\mathrm{SU}(a/Eb_n) = n$, which implies that $\mathrm{SU}(a/E) \geq \omega$. On the other hand, any forking extension of $\mathrm{tp}(a/E)$ has finite SU-rank (since if $a$ is not independent from $F$ over $E$, then $a$ satisfies some $\sigma$-equation, i.e., $\mathrm{tr.deg}(\mathrm{cl}_\sigma(Fa)/F) < \infty$). Hence, $\mathrm{SU}(a/E) = \omega$.

Note that

$$\omega = \mathrm{SU}(a, b_1/E)$$
$$= \mathrm{SU}(a/Eb_1) + \mathrm{SU}(b_1/E) < \mathrm{SU}(a/Eb_1) \oplus \mathrm{SU}(b_1/E) = \omega + 1.$$

**4.18. Other examples.** Let $V$ be a variety of dimension $d$ defined over some $E = \mathrm{acl}_\sigma(E) \subseteq K$. Assume that if $\bar{a} = (a_1, \ldots, a_n)$ is a tuple of $V(K)$ which is generic over $E$, then the elements $a_1, \ldots, a_d$ are algebraically independent over $E$. Consider the type $p_V(\bar{x})$ over $E$ which says that $x_1, \ldots, x_d$ are transformally independent over $E$ (i.e., satisfy no nontrivial difference equation over $E$), and that $(x_1, \ldots, x_n) \in V$. According to the additivity rule (one can also see it directly), if $\bar{a}$ realizes $p_V$, then $\mathrm{SU}(\bar{a}/E) = \omega d$. The type $p_V(\bar{x})$ is, in a sense, a *generic type of the variety* $V$. One can show that this type is complete.

Finally, one shows easily that if $\bar{a}$ is any tuple of $K$, then $\mathrm{SU}(\bar{a}/E)$ is of the form $\omega n + m$ for some nonnegative integers $n, m$: if $\bar{b} \subset \bar{a}$ is maximal transformally independent over $E$ and has length $n$, and if $m = \mathrm{tr.deg}(\mathrm{cl}_\sigma(E\bar{a})/\mathrm{cl}_\sigma(E\bar{b}))$, then $\omega n \leq SU(\bar{a}/E) \leq \omega n + m$.

**4.19. Definition of imaginaries.** Let $M$ be a model of a complete theory $T$ in a language $\mathcal{L}$. We assume $M$ to be sufficiently saturated. Let $S \subseteq M^n$ be a 0-definable set, and let $E \subseteq S^2$ be a 0-definable equivalence relation. Then each $E$-equivalence class $\bar{a}/E$, with $\bar{a} \in S$, is called an *imaginary* element.

Given $S$ and $E$ as above, the set $S/E$ is interpretable in $M$. To each such pair, we associate a new sort, and let $M^{\mathrm{eq}}$ be the many-sorted structure with sorts indexed by the pairs $(S, E)$ as above, the "real sort" being the $\mathcal{L}$-structure $M$, and the sort indexed by $(S, E)$ being the set $S/E$; there is also a projection map $S \to S/E$ for each $(S, E)$. The structure $M^{\mathrm{eq}}$ is then interpretable in $M$. The difference between many-sorted logic and 1-sorted logic is that all quantifiers, variables, and constants have a sort attached to them so that you will have things like $\forall x \in M$, $\forall y \in S/E$. Note also that $M^n$ becomes a sort so that an $n$-tuple of $M$ can be thought of as an element of $M^{\mathrm{eq}}$.

An important example of imaginary is the following. Let $D \subseteq M^k$ be a definable set, defined by some formula $\varphi(\bar{x}, \bar{b})$, where $\varphi(\bar{x}, \bar{y}) \in \mathcal{L}$ and $\bar{b}$ is an

$n$-tuple from $M$. Define an equivalence relation $E$ on $M^n$ by

$$E(\bar{y}, \bar{z}) \colon \forall \bar{x} \, (\varphi(\bar{x}, \bar{y}) \leftrightarrow \varphi(\bar{x}, \bar{z})).$$

Then the equivalence class $\bar{b}/E$ has the following property: for all $\rho \in \mathrm{Aut}(M)$, $\rho(D) = D$ if and only if $\rho$ fixes $\bar{b}/E$.

The imaginary element $\bar{b}/E$ will be called a *code for $D$*.

We say that $T$ *eliminates imaginaries* if whenever $D \subseteq M^k$ is definable (with parameters), then there is a tuple $\bar{d}$ in $M$ such that for any $\rho \in \mathrm{Aut}(M)$,

$$\rho(D) = D \iff \rho \text{ fixes the elements of the tuple } \bar{d}.$$

If the language $\mathcal{L}$ has at least two terms $t_0, t_1$ and $T \models t_0 \neq t_1$, then $T$ eliminates imaginaries if and only if whenever $S \subseteq M^n$ is 0-definable, and $E \subseteq S^2$ is a 0-definable equivalence relation, then there is a 0-definable function $f \colon S \to M^k$ for some $k$ such that for every $\bar{y}, \bar{z} \in S$ we have

$$M \models f(\bar{y}) = f(\bar{z}) \iff E(\bar{y}, \bar{z}).$$

**4.20. Some facts.** (1) If $T$ eliminates imaginaries, and $A \subset M$, then $T(A)$ eliminates imaginaries as well ($T(A)$ is the set of sentences in the language $\mathcal{L}(A)$, obtained by adjoining to $\mathcal{L}$ a constant symbol for each element of $A$, which hold in the $\mathcal{L}(A)$-structure $(M, a)_{a \in A}$.)

(2) □□□ Assume that $M$ is stable. Working in $M^{\mathrm{eq}}$ instead of $M$, one has the following: if $a$ is independent from $b$ over $cd$ and from $c$ over $bd$, then $a$ is independent from $(bc)$ over $\mathrm{acl}^{\mathrm{eq}}(bd) \cap \mathrm{acl}^{\mathrm{eq}}(cd)$. Note that a tuple of elements of $M$ can be thought of as an imaginary element. The proof uses the following ingredients. By elimination of imaginaries and stability, we know that every type $p$ over an algebraically closed set $E$ is *stationary* (i.e., has a unique nonforking extension to any superset of $E$) and is *definable over $E$*, i.e., given a formula $\varphi(\bar{x}, \bar{y}) \in \mathcal{L}$, there is a formula $d_\varphi(\bar{y}) \in \mathcal{L}(E)$ such that for every tuple $\bar{b}$ in $E$, we have that

$$\varphi(\bar{x}, \bar{b}) \in p \iff M \models f_\varphi(\bar{b}).$$

The nonforking extension $p'$ of $p$ to a set $F$ containing $E$ will then be defined analogously: for every $\bar{b} \in F$, $\varphi(\bar{x}, \bar{b}) \in p' \iff M \models d_\varphi(\bar{b})$. One defines the *canonical base of $p$*, denoted by $\mathrm{Cb}(p)$, to be the set of codes of the formulas $d_\varphi(\bar{y})$ (and by elimination of imaginaries, this is a subset of $E$). By definition we have that $p$ does not fork over $\mathrm{Cb}(p)$, and that the restriction of $p$ to $\mathrm{Cb}(p)$ is stationary. Moreover, if $E_0 \subset E$ is such that $p$ does not fork over $E_0$, then $\mathrm{acl}(E_0)$ contains $\mathrm{Cb}(p)$.

Hence, we look at the canonical base of $\mathrm{tp}(a/\mathrm{acl}(bcd))$ and from the independence relations, deduce that it is contained in $\mathrm{acl}(bd) \cap \mathrm{acl}(cd)$.

We will use the fact that every completion of the theory ACF of algebraically closed fields eliminates imaginaries and is stable. Note that the above proof extends to the case of $M$ simple with stable forking.                                        □□□

(2′) Let us rephrase (2) in a more algebraic language. We work within a sufficiently large algebraically closed field $\Omega$. It is known that any algebraic set $V$ defined over $\Omega$ has a *smallest field of definition*. If one looks at what it means in terms of independence, it translates as follows: let $E$ and $F$ be algebraically closed subfields of $\Omega$, let $\bar{a}$ be a tuple of elements of $\Omega$, and assume that

$$\bar{a} \underset{E}{\downarrow} F \quad \text{and} \quad \bar{a} \underset{F}{\downarrow} E.$$

This means that if $V$ is the locus of $\bar{a}$ over $(EF)^{\mathrm{alg}}$, then $V$ is defined over $E$ and $V$ is defined over $F$. Indeed, let $V_1$ be the locus of $\bar{a}$ over $E$. As $E \subset EF$, we certainly have $V_1 \supseteq V$. From the fact that $\bar{a}$ is free from $F$ over $E$, we know that the dimensions of $V_1$ and $V$ (which equal, respectively, tr.deg$(\bar{a}/E)$ and tr.deg$(\bar{a}/EF)$) are equal. As $V_1$ is irreducible, this implies that $V = V_1$ so that $V$ is defined over $E$. Reasoning similarly with the locus of $\bar{a}$ over $F$, one obtains that $V$ is defined over $F$.

The uniqueness of the smallest field of definition of $V$ then implies that $V$ is defined over $E \cap F$. Hence, all equations over $EF$ satisfied by $\bar{a}$ are implied by equations over $E \cap F$. This implies that tr.deg$(\bar{a}/E \cap F) = \dim(V)$ so that

$$\bar{a} \underset{E \cap F}{\downarrow} EF.$$

(3) In the case of fields, any finite set has a code. Let us show how it works for elements of $M$: we want to code the definable set $\{a_1, \ldots, a_n\}$. Consider the polynomial $f(X) = \prod_{i=1}^{n}(X - a_i)$, and let $b_1, \ldots, b_n$ be its coefficients. Then any permutation of $\{a_1, \ldots, a_n\}$ fixes $b_1, \ldots, b_n$, and the formula $f(x) = 0$ defines $\{a_1, \ldots, a_n\}$. Hence the tuple $(b_1, \ldots, b_n)$ is a code for the set $\{a_1, \ldots, a_n\}$. There is a similar trick for finite sets of tuples.

(4) The uniqueness of the field of definition of algebraic sets also gives easily that any completion of the theory of algebraically closed fields eliminates imaginaries. Indeed, let $K$ be an algebraically closed field. By elimination of quantifiers, we know that any definable subset $D$ of $K^n$ is a Boolean combination of algebraic sets. Then $D$ can be written $V \setminus W$, where $V$ is the Zariski closure of $D$ and $W$ is some definable subset of $V$ with the property that every irreducible component of its Zariski closure $\bar{W}$ is strictly contained in some irreducible component of $V$. In particular, $\dim(\bar{W}) < \dim(V)$. Clearly any automorphism $\rho$ of $K$ which leaves $D$ invariant will also leave $V$ and $W$ invariant. The automorphism $\rho$ leaves $V$ invariant if and only if it fixes the field of definition of $V$. The result follows by induction on the dimension.

This type of proof generalizes to other theories of fields which eliminate quantifiers: the theory of differentially closed fields of characteristic 0, and also the theory of separably closed fields of finite degree of imperfection (in that case, one needs, however, to add some constant symbols to the language).

**4.21. Theorem.** Any completion of ACFA eliminates imaginaries.

SKETCH OF THE PROOF □□□ . We work in a saturated model $K$ of ACFA. We are given a 0-definable function $f$ and a tuple $\bar{a}$, and we look at the equivalence class $e$ of $\bar{a}$ for the equivalence relation $E(\bar{x}, \bar{y}) \iff f(\bar{x}) = f(\bar{y})$. We want to show that there is a real tuple which is equidefinable with $e$.

Let $E = \mathrm{acl}^{\mathrm{eq}}(e) \cap K$. If $e$ is definable over $E$, then we are done: choose a tuple $\bar{b} \in E$ over which $e$ is definable. Then $\bar{b} \in \mathrm{acl}^{\mathrm{eq}}(e)$. Since we are in a field, there is a tuple $\bar{c}$ which codes the finite set of conjugates of $\bar{b}$ over $e$. Then $\bar{c}$ and $e$ are equidefinable. Hence, we may assume that $e$ is not definable over $E$, and in particular that $\bar{a} \notin E$. Our aim is to show that there is a tuple $\bar{b}$ realizing $\mathrm{tp}(\bar{a}/e)$ which is independent from $\bar{a}$ over $E$.

Let $p = \mathrm{tp}(\bar{a}/e)$. Since $p$ is nonalgebraic, there is a $\bar{b}$ realizing $p$ and such that $\mathrm{acl}^{\mathrm{eq}}(e\bar{a}) \cap \mathrm{acl}^{\mathrm{eq}}(e\bar{b}) = \mathrm{acl}^{\mathrm{eq}}(e)$, and therefore,

$$\mathrm{acl}_\sigma(E\bar{a}) \cap \mathrm{acl}_\sigma(E\bar{b}) = E. \qquad (*)$$

Choose $\bar{b}$ realizing $p$, satisfying $(*)$ and of maximal SU-rank over $E\bar{a}$ (note that such a $\bar{b}$ exists, by the rank inequality given in Subsection 4.18, and that $f(\bar{b}) = e$). Let $\bar{c}$ realize $\mathrm{tp}(\bar{b}/E\bar{a})$ independent from $\bar{b}$ over $E\bar{a}$. Then $c$ realizes $p$, as $e \in \mathrm{dcl}^{\mathrm{eq}}(\bar{a})$. Moreover, $\mathrm{acl}_\sigma(E\bar{c}) \cap \mathrm{acl}_\sigma(E\bar{b}) \subseteq \mathrm{acl}_\sigma(E\bar{b}) \cap \mathrm{acl}_\sigma(E\bar{a})$ (because $\mathrm{acl}_\sigma(E\bar{a}\bar{b}) \cap \mathrm{acl}_\sigma(E\bar{a}\bar{c}) = \mathrm{acl}_\sigma(E\bar{a})$), and therefore the tuple $(\bar{b}, \bar{c})$ also satisfies $(*)$. By maximality of the SU-rank of $\bar{b}$ over $E\bar{a}$ and because $\mathrm{tp}(\bar{a}/e) = \mathrm{tp}(\bar{b}/e)$, we get that

$$\mathrm{SU}(\bar{c}/E\bar{b}) \leq \mathrm{SU}(\bar{b}/E\bar{a}).$$

We also know that

$$\mathrm{SU}(\bar{c}/E\bar{a}\bar{b}) = \mathrm{SU}(\bar{c}/E\bar{a}) = \mathrm{SU}(\bar{b}/E\bar{a}).$$

As $\mathrm{SU}(\bar{c}/E\bar{a}\bar{b}) \leq \mathrm{SU}(\bar{c}/E\bar{b})$, we obtain that

$$\mathrm{SU}(c/E\bar{a}\bar{b}) = \mathrm{SU}(\bar{b}/E\bar{a}) = \mathrm{SU}(\bar{c}/E\bar{b}),$$

so that $\mathrm{cl}_\sigma(\bar{c})$ is independent from $\mathrm{acl}_\sigma(E\bar{a}\bar{b})$ over $\mathrm{acl}_\sigma(E\bar{a})$ and over $\mathrm{acl}_\sigma(E\bar{b})$ (in the sense of the theory of algebraically closed fields). By elimination of imaginaries of the theory of algebraically closed fields, this implies that $\mathrm{cl}_\sigma(\bar{c})$ is (ACF-)independent from $\mathrm{cl}_\sigma(\bar{a}, \bar{b})$ over $\mathrm{acl}_\sigma(E\bar{a}) \cap \mathrm{acl}_\sigma(E\bar{b}) = E$, i.e., that $\bar{c}$ is (ACFA-)independent from $(\bar{a}, \bar{b})$ over $E$. Hence $\mathrm{SU}(\bar{b}/E\bar{a}) = \mathrm{SU}(\bar{c}/E\bar{a}) = \mathrm{SU}(\bar{c}/E)$ so that $\bar{a}$ and $\bar{b}$ are independent over $E$.

Hence, we have shown that there is a realization $\bar{b}$ of $p$ which is independent from $\bar{a}$ over $E$. Since $e$ is not definable over $E$, there is $\bar{a}'$ realizing $\mathrm{tp}(\bar{a}/E)$ and with $f(\bar{a}') \neq f(\bar{a})$. Since $\mathrm{tp}(\bar{a}'/E) = \mathrm{tp}(\bar{a}/E)$, there is $\bar{c}'$ realizing $\mathrm{tp}(\bar{a}'/E)$, with $f(\bar{a}') = f(\bar{c}')$ and independent from $\bar{a}'$ over $E$ and we may assume that this $\bar{c}'$ is also independent from $\bar{b}$ over $E$. Apply the independence theorem to $\mathrm{tp}(\bar{a}/E\bar{b}) \cup \mathrm{tp}(\bar{a}'/E\bar{c}')$ to derive a contradiction. ⊣ □□□

**4.22. Corollary.** Let $\bar{a}, \bar{b}, \bar{c}, \bar{d}$ be tuples in $K$, and assume that $\bar{a}$ is independent from $\bar{b}$ over $(\bar{c}\bar{d})$ and from $\bar{c}$ over $(\bar{b}\bar{d})$. Then $\bar{a}$ is independent from $(\bar{b}\bar{c})$ over $\mathrm{acl}_\sigma(\bar{b}\bar{c}) \cap \mathrm{acl}_\sigma(\bar{b}\bar{d})$.

PROOF. Either use the remark about simple theories with stable forking given in Subsection 4.20(2), or equivalently, the fact that independence in models of ACFA corresponds to independence of algebraically closed sets for the theory of algebraically closed fields. One uses also the fact that $\mathrm{acl}_\sigma(AB) = (\mathrm{acl}_\sigma(A)\mathrm{acl}_\sigma(B))^{\mathrm{alg}}$. ⊣

**4.23. A useful remark.** Let $K$ be a model of ACFA, $E$ an algebraically closed subset of $K$, and $\bar{a}$ a tuple of elements of $K$. As in the case of algebraically closed fields, one can show that $I_\sigma(\bar{a}/E)$ has a smallest field of definition (as a $\sigma$-ideal). Or more simply, this follows from the elimination of imaginaries. So, let us suppose that $E$ is the algebraic closure of this smallest field of definition, so that, in particular, it is the algebraic closure of a *finite tuple* and therefore is ranked by the SU-rank.

CLAIM. For $n$ sufficiently large, if $\bar{a}_0, \ldots, \bar{a}_n$ are independent realizations of $\mathrm{tp}(\bar{a}/E)$, then $E \subseteq \mathrm{acl}_\sigma(\bar{a}_0, \ldots, \bar{a}_n)$.

PROOF. We assume $K$ sufficiently saturated. Construct by induction on $n \in \mathbb{N}$ a sequence $\bar{a}_n$, $n \in \mathbb{N}$, of realizations of $\mathrm{tp}(\bar{a}/E)$, with $\bar{a}_n \underset{E}{\overset{}{\downarrow}} \bar{a}_0, \ldots, \bar{a}_{n-1}$ for every $n$. Then

$$\mathrm{SU}(E/\bar{a}_0, \ldots, \bar{a}_n) \leq \mathrm{SU}(E/\bar{a}_0, \ldots, \bar{a}_{n-1})$$

for every $n$, so that there is some index $m$ such that $\mathrm{SU}(E/\bar{a}_0, \ldots, \bar{a}_{m+1}) = \mathrm{SU}(E/\bar{a}_0, \ldots, \bar{a}_m)$. Take the smallest such $m$. Then

$$E \underset{\bar{a}_0, \ldots, \bar{a}_m}{\overset{}{\downarrow}} \bar{a}_{m+1},$$

and by assumption

$$\bar{a}_{m+1} \underset{E}{\overset{}{\downarrow}} \bar{a}_0, \ldots, \bar{a}_m.$$

Corollary 4.22 then gives

$$\bar{a}_{m+1} \underset{E \cap \mathrm{acl}_\sigma(\bar{a}_0, \ldots, \bar{a}_m)}{\overset{}{\downarrow}} E, \bar{a}_0, \ldots, \bar{a}_m,$$

and, therefore, $E \subseteq \mathrm{acl}_\sigma(\bar{a}_0, \ldots, \bar{a}_m)$. ⊣

**4.24. Exercise 8.** Show that the formula $\sigma^2(x) = x^2$ has SU-rank 1. That is, you need to show that given any algebraically closed difference field $E$ and element $a$ satisfying $\sigma^2(x) = x^2$, one *cannot have* $\mathrm{tr.deg}(\mathrm{cl}_\sigma(Ea)/E) = 1$.

**§5. Study of the fixed field.** A particularly important definable subset of a model $K$ of ACFA is the fixed field, $\mathrm{Fix}(\sigma) = \{x \in K \mid \sigma(x) = x\}$. It turns out that this subfield is responsible for much of the bad behavior of models of ACFA. We will show that $\mathrm{Fix}(\sigma)$ is a pseudofinite field, i.e., is elementarily equivalent to an ultraproduct of finite fields (or equivalently, is an infinite model of the theory of finite fields). Pseudofinite fields were first studied by Ax, see [9]. Some very nice results on pseudofinite fields were also obtained in two papers by E. Hrushovski and A. Pillay ([11] and [12]).

**5.1. Theorem.** Let $K$ be a model of ACFA, and $F = \mathrm{Fix}(\sigma)$. Then $F$ is a pseudofinite field.

PROOF. We need to show that

(i) $F$ is perfect (i.e., if $\mathrm{char}(F) = p > 0$, then every element is a $p$th power).

(ii) $\mathcal{G}\mathrm{al}(F^{\mathrm{alg}}/F) \simeq \hat{\mathbb{Z}} \simeq \lim_{\leftarrow} \mathbb{Z}/n\mathbb{Z} \simeq \prod_{p \text{ prime}} \mathbb{Z}_p$.

(iii) $F$ is pseudoalgebraically closed (PAC), i.e., every variety defined over $F$ has a point with coordinates in $F$.

Item (i) is no problem: if the characteristic is $p > 0$, then every element $a$ has a unique $p$th root, denoted by $a^{1/p}$. Hence $\sigma(a) = a$ implies $\sigma(a^{1/p}) = a^{1/p}$, and $F$ is perfect. Item (iii) is no problem either: let $U$ be a variety defined over $F$, and consider the diagonal subvariety $V \subseteq U \times U$, i.e., $V$ is defined by $\bar{x} \in U$, $\bar{y} \in U$, and $\bar{x} = \bar{y}$. Then $U = U^\sigma$, and $U, V$ satisfy the hypotheses of axiom (3) of ACFA, so that there is $\bar{a} \in K$ with $(\bar{a}, \sigma(\bar{a})) \in V$, i.e., $\bar{a} \in U$ and $\sigma(\bar{a}) = \bar{a}$.

Let us now look at item (ii). First of all, if $L$ is a finite Galois extension of $F$, then $\sigma(L) = L$: if $\alpha$ generates $L$ over $F$, then $\sigma$ fixes the coefficients of the minimal monic polynomial of $\alpha$ over $F$, so that $\sigma(\alpha) \in L$. Hence $\sigma$ restricts to an element of $\mathcal{G}\mathrm{al}(L/F)$. By Galois theory, $F = \mathrm{Fix}(\sigma)$ is the subfield of $L$ fixed by the group generated by $\sigma|_L$, and this shows that $\mathcal{G}\mathrm{al}(L/F)$ is cyclic, generated by $\sigma|_L$.

We will now show that for every $n$, $F$ has *at most one* Galois extension of degree $n$. Given a finite Galois extension $L$ of $F$, there is a 1-1 correspondence between algebraic extensions of $F$ contained in $L$ and subgroups of $\mathcal{G}\mathrm{al}(L/F)$, under which the extensions which are Galois over $F$ correspond to the normal subgroups of $\mathcal{G}\mathrm{al}(L/F)$. Every finite algebraic extension of $F$ is contained in a finite Galois extension of $F$, and all subgroups of a cyclic group are normal. Hence every algebraic extension of $F$ is Galois.

Assume now by way of contradiction that $F$ has two Galois extensions, $L$ and $M$, of the same degree $n$ over $F$. Consider $\mathcal{G}\mathrm{al}(LM/F)$. Then $\mathcal{G}\mathrm{al}(LM/L)$ and $\mathcal{G}\mathrm{al}(LM/M)$ are subgroups of $\mathcal{G}\mathrm{al}(LM/F)$ of order $d = [LM : F]/n$. But $\mathcal{G}\mathrm{al}(LM/F)$ is cyclic, and therefore has only one subgroup of order $d$: this implies that $L = M$.

To show (ii), it will be enough to show that for every $n$, $F$ has *at least* one Galois extension of degree $n$. But this is easy: consider the difference field extension $L = K(X_1, \ldots, X_n)$ of $K(X_1, \ldots, X_n$ indeterminates) with $\sigma(X_i) = X_{i+1}$ for $i = 1, \ldots, n-1$, and $\sigma(X_n) = X_1$. Then

$$L \models \exists x\, \sigma^n(x) = x \wedge \bigwedge_{1 \leq i < n} \sigma^i(x) \neq x,$$

so that $K$ satisfies the same sentence. Let $a \in K$ be such that $\sigma^n(a) = a$, $\sigma^i(a) \neq a$ for $1 \leq i < n$. By Galois theory, $a$ generates a Galois extension of degree $n$ of $F$. ⊣

**5.2. An example.** We saw earlier that if $A$ is an algebraically closed difference field, then $ACFA \cup qftp(A)$ is complete. Hence, in a sense, ACFA is close to having quantifier-elimination. However, it does not. Here is an example.

Let $a, b \in Fix(\sigma)$ be transcendental elements, and assume that $a$ does not have a square root in $Fix(\sigma)$ (so we assume that $char(Fix(\sigma)) \neq 2$). Then we have

$$K \models \forall y \; y^2 = a \rightarrow \sigma(y) \neq y$$
$$K \models \exists y \; y^2 = b^2 \wedge \sigma(y) = y.$$

However, $a$ and $b^2$ have the same quantifier-free type, as the quantifier-free type of a transcendental element of the fixed field is unique: it simply says that the element satisfies no nontrivial equation (over $\mathbb{Q}$ or over $\mathbb{F}_p$).

**5.3. Proposition.** Let $K \models ACFA$, and consider $F = Fix(\sigma) = \{x \in K \mid \sigma(x) = x\}$. Every definable subset of $F^n$ is definable in the pure field $F$. In other words, the structure on $F$ induced by $K$ is the one of the pure field $F$.

PROOF. We assume $K$ sufficiently saturated. Let $S \subset F^n$ be definable (in $K$). Then $\sigma(S) = S$. By elimination of imaginaries, this implies that there is a tuple $\bar{c}$ of elements of $F$, and a formula $\varphi(\bar{x}, \bar{y})$ such that $\varphi(\bar{x}, \bar{c})$ defines $S$. Hence, we have shown that $S$ is definable over $F$. To finish the proof, it suffices to show that there is a "small" subset $C$ of $F$ such that every field-automorphism of $F$ which fixes $C$ is an elementary map within the structure $K$ (exercise).

Let $C$ be a countable elementary substructure of the field $F$. Then $C^{alg}$ and $F$ are linearly disjoint over $C$ because $C \prec F$. If $C_n$ is the unique algebraic extension of $C$ of degree $n$, then $FC_n$ is an algebraic extension of $F$ of degree $n$ also, and therefore is the unique algebraic extension of $F$ of degree $n$. Hence $F^{alg} = FC^{alg}$.

Let $\rho$ be a field-automorphism of $F$ which is the identity on $C$. Because $C^{alg}$ and $F$ are linearly disjoint over $C$, we can extend $\rho$ to a field-automorphism $\tilde{\rho}$ of $FC^{alg}$ which is the identity on $C^{alg}$. Since $F^{alg} = FC^{alg}$, we get that $\tilde{\rho}$ is defined on $F^{alg} = acl_\sigma(F)$. We have

$$\sigma\tilde{\rho}\big|_F = \tilde{\rho}\sigma\big|_F$$

because $\sigma\big|_F = id_F$, and

$$\sigma\tilde{\rho}\big|_{C^{alg}} = \tilde{\rho}\sigma\big|_{C^{alg}}$$

because $\rho\big|_{C^{alg}} = id_{C^{alg}}$. Hence, $\tilde{\rho}$ commutes with $\sigma$ on $C^{alg}$ and on $F$, so that $\tilde{\rho}$ is an automorphism of the difference field $acl_\sigma(F)$. By Corollary 3.15, $\tilde{\rho}$ is an elementary map. ⊣

**5.4. Exercise 9.** Let $K$ be a model of ACFA, $F$ its fixed field.

(1) Let $E$ be a subfield of $F$ such that $E^{\mathrm{alg}} F = F^{\mathrm{alg}}$, and $\bar{a}$ a tuple of elements of $F$. Show that $\mathrm{SU}(\bar{a}/E) = \mathrm{tr.deg}(E(\bar{a})/E)$.
(2) Let $S$ be a definable subset of $F^n$. Show that $\mathrm{SU}(S)$ equals the algebraic dimension of the Zariski closure (in $K$) of $S$.

**§6. Orthogonality and modularity.** In this section we will show that we can reduce the study of types of finite SU-rank to the study of types of SU-rank 1. We will always be working in a model $K$ of ACFA which we will assume to be sufficiently saturated.

**6.1. Definitions.** Let $A$ and $B$ be subsets of $K$, and $p, q$ types over $A$ and $B$, respectively.

(1) If $A = B$, then we say that $p$ and $q$ are *almost orthogonal* (denoted by $p \perp^a q$) if whenever $\bar{a}$ realizes $p$ and $\bar{b}$ realizes $q$, then $\bar{a}$ and $\bar{b}$ are independent over $A$.
(2) We say that $p$ and $q$ are *orthogonal* (denoted by $p \perp q$) if whenever $C$ contains $A \cup B$, $\bar{a}$ is a realization of $p$ which is independent from $C$ over $A$, and $\bar{b}$ is a realization of $q$ which is independent from $C$ over $B$, then $\bar{a}$ and $\bar{b}$ are independent over $C$.
(3) Recall that $S \subseteq K^n$ is $\infty$-*definable over* $E$, or *type-definable over* $E$, if there is a partial $n$-type $\Phi$ over $E$ (i.e., a consistent set of $\mathcal{L}(E)$-formulas in the variables $(x_1, \ldots, x_n)$) such that $S$ is the set of $n$-tuples from $K$ satisfying $\Phi$.
(4) Let $S$ be a set which is ($\infty$-) definable over $B$. We say that $p$ is *orthogonal to* $S$ (denoted by $p \perp S$, or by $p \perp \varphi$ if $S$ is defined by $\varphi$) if $p$ is orthogonal to all types over supersets of $B$ which are realized in $S$; i.e., for all $C$ containing $A \cup B$, and $\bar{a}$ realizing $p$ and independent from $C$ over $A$, and $\bar{b} \in S$, we have that $\bar{a}$ and $\bar{b}$ are independent over $C$. One also says that $p$ is *foreign* to $S$ or to $\varphi$.

**6.2. Remark and example.** Note that in (4) above, we do not require that $\bar{b}$ be independent from $C$ over $B$. Here is an example which shows that orthogonality to a set and to a type are different.

Let $E = \mathrm{acl}_\sigma(E) \subseteq K$. Let $p$ be the type of a transformally transcendental element $a$ over $E$. Let $P$ be the set of realizations of $p$ over $E$.

Then $p$ is orthogonal to all types of finite SU-rank over $E$: indeed, assume that $a$ is transformally transcendental over $F = \mathrm{acl}_\sigma(F) \supset E$, and that $\mathrm{SU}(b/F) < \omega$, but $a$ and $b$ are not independent over $F$. This means that $I_\sigma(a/\mathrm{acl}_\sigma(Fb))$ is nonzero, and therefore that $a$ is transformally algebraic over $\mathrm{acl}_\sigma(Fb)$, i.e., $\mathrm{SU}(a/Fb) < \omega$. But this contradicts Lascar's inequality: on the one hand, we have $\mathrm{SU}(a, b/F) \geq \mathrm{SU}(a/F) = \omega$, and on the other, we have $\mathrm{SU}(a, b/F) \leq \mathrm{SU}(a/Fb) \oplus \mathrm{SU}(b/F) < \omega$.

However, let $c = \sigma(a) - a$, let $a'$ be a realization of $\text{tp}(a/\text{acl}_\sigma(Ec))$, independent from $a$ over $Ec$, and let $b = a - a'$. Then $\sigma(b) = \sigma(a) - \sigma(a') = (a + c) - (a' + c) = a - a' = b$ so that $\text{SU}(b/E) = 1 = \text{SU}(b/Ea')$. So we have that $\text{tp}(b/E) \not\perp \text{tp}(a/Ea')$, and therefore $\text{tp}(b/E) \not\perp P$.

**6.3. Another comment about orthogonality.** Orthogonality or nonorthogonality tells us about interactions between sets. For instance, let $D_1$ and $D_2$ be two infinite definable sets, defined over some $E = \text{acl}_\sigma(E)$.

If all types realized in $D_1$ are orthogonal to all types realized in $D_2$, then this means that whenever you take a tuple $\bar{a}$ of elements in $D_1$ and a tuple $\bar{b}$ of elements of $D_2$, then they are independent over any set containing $E$. In particular, any definable map $f: S \to D_2^m$, where $S \subseteq D_1^n$, for some integers $m, n$, will take only finitely many values.

Assume in addition that all types realized in $D_1$ and in $D_2$ are stationary (recall that a type over a set $A$ is *stationary* if it has a unique nonforking extension to any set containing $A$). It will then follow that any definable subset of $D_1 \times D_2$ is a finite union of rectangles, i.e., is of the form $\bigcup_{i=1}^n S_i \times T_i$, where $S_i$ is a definable subset of $D_1$ and $T_i$ is a definable subset of $D_2$.

**6.4. Exercise 10.** Let $E = \text{acl}_\sigma(E) \subset K$, $\bar{a}, \bar{b}$ tuples from $K$, and assume that $\text{SU}(\bar{a}/E) = \text{SU}(\bar{b}/E) = 1$. Show that $\text{tp}(\bar{a}/E) \not\perp \text{tp}(\bar{b}/E)$ implies that $\text{tr.deg}(\text{cl}_\sigma(E\bar{a})/E) = \text{tr.deg}(\text{cl}_\sigma(E\bar{b})/E)$.

**6.5. Exercise 11.** Let $p, q, r$ be types over algebraic closed sets $A, B$, and $C$, respectively. Assume that $p, q$, and $r$ have SU-rank 1, and that $A \mathrel{\underset{B}{\smile}} C$. Show that if $p \not\perp q$ and $q \not\perp r$, then $p \not\perp r$. (Warning: your proof should use all the assumptions on $A, B, C$: their being algebraically closed and the independence hypothesis. Nonorthogonality *is not* an equivalence relation on types of SU-rank 1. One can, however, show that the relation $E$ defined by $E(p, q)$ if and only if there is $r$ of SU-rank 1 such that $p \not\perp r$ and $q \not\perp r$ defines an equivalence relation on types of SU-rank 1.)

**6.6. Exercise 12.** Let $E = \text{acl}_\sigma(E) \subset K$, and $a \in K$ with $\text{SU}(a/E) = 1$. Assume that $\text{tp}(a/E) \not\perp (\sigma(x) = x)$. We know that $\text{tr.deg}(\text{cl}_\sigma(Ea)/E) = 1$, so that $E(a)^{\text{alg}} = \text{acl}_\sigma(Ea)$. Show that there exists an integer $N$ such that $[E(a, \sigma^k(a)) : E(a)] \leq N$ for every $k \in \mathbb{Z}$. (Hint: take $E' = \text{acl}_\sigma(E')$ independent from $a$ over $E$ such that there is some $b \in \text{acl}_\sigma(E'a) \backslash E'$ with $\sigma(b) = b$. Then for every $k$ one has $[E(a, \sigma^k(a)) : E(a)] = [E'(a, \sigma^k(a)) : E'(a)]$, so we may assume that $E = E'$. Let $m = [E(a, b) : E(a)]$, $n = [E(a, b) : E(b)]$. Show that $[E(a, b, \sigma^k(a)) : E(a)] \leq mn$.)

**6.7. Remark.** One can show that the converse is true, but the proof is much harder.

**6.8. Exercise 13/Example.** Let $E = \text{acl}_\sigma(E) \subset K$, let $a \in E$, and let $b \in K \setminus E$ satisfy $\sigma(x) - x = a$.

(a) Show that $\text{tp}(b/E) \not\perp (\sigma(x) = x)$.

(b) Show that $\text{tp}(b/E)$ is not almost orthogonal to some type containing $\sigma(x) = x$, if and only if there is an integer $m > 0$ and $\alpha \in E$ satisfying $\sigma^m(x) = x + a + \cdots + \sigma^{m-1}(a)$. (Hint: assume that $c \in (\text{acl}_\sigma(Eb) \setminus E) \cap \text{Fix}(\sigma)$; looking at the coefficients of the minimal polynomial of $c$ over $\text{cl}_\sigma(Eb)$, we may in fact assume that $c \in \text{cl}_\sigma(Eb)$, so that $c = g(b)$ for some $g(X) \in E(X)$. Look at the sets $S_0$ of poles and $S_1$ of zeroes of $g$, and use the equation $g(b) = g^\sigma(b + a)$.)

**6.9. Proposition.** Every finite SU-rank type is nonorthogonal to a type of SU-rank 1.

PROOF. Let $E = \text{acl}_\sigma(E) \subset K$, and let $\bar{a}$ be a tuple in $K$ with $\text{SU}(\bar{a}/E) = n < \omega$. We want to find $F = \text{acl}_\sigma(F)$ independent from $\bar{a}$ over $E$, and $b \in \text{acl}_\sigma(F\bar{a}) \setminus F$ such that $\text{SU}(b/F) = 1$.

By definition of the SU-rank, there is some tuple $\bar{d}$ such that $\text{SU}(\bar{a}/E\bar{d}) = n - 1$. Given such a tuple, we may always write it as $(\bar{b}, \bar{c})$, with $\bar{c}$ independent from $\bar{a}$ over $E$. Find $(\bar{b}, \bar{c})$ such that $\bar{c}$ and $\bar{a}$ are independent over $E$, $\text{SU}(\bar{a}/E\bar{b}\bar{c}) = n - 1$, and $\text{SU}(\bar{b}/E\bar{c})$ is least possible.

Let $\bar{a}'$ realize $\text{tp}(\bar{a}/E\bar{b}\bar{c})$ and independent from $\bar{a}$ over $E\bar{b}\bar{c}$. Then $\text{SU}(\bar{a}/E\bar{b}\bar{c}\bar{a}') = \text{SU}(\bar{a}/E\bar{b}\bar{c}) = n - 1$.

CLAIM. $\bar{a}$ and $\bar{a}'$ are not independent over $E\bar{c}$.

PROOF OF CLAIM. Otherwise, assume that $\bar{a}$ and $\bar{a}'$ are independent over $E\bar{c}$, and let $\bar{c}' = (\bar{c}, \bar{a}')$. Since $\bar{a}'$ realizes $\text{tp}(\bar{a}/E\bar{b}\bar{c})$, we have $\text{SU}(\bar{a}'/E\bar{b}\bar{c}) < \text{SU}(\bar{a}'/E\bar{c})$; hence, by symmetry we get $\text{SU}(\bar{b}/E\bar{c}\bar{a}') < \text{SU}(\bar{b}/E\bar{c})$, so that the pair $(\bar{b}, \bar{c}')$ contradicts the minimality of $\text{SU}(\bar{b}/E\bar{c})$.  ⊣

Hence $\bar{a}'$ and $\bar{a}$ are not independent over $E\bar{c}$, so that $\text{SU}(\bar{a}/E\bar{c}\bar{a}') = \text{SU}(\bar{a}'/E\bar{c}\bar{a}) = n - 1$. It follows that $\bar{a}'$ is independent from $\bar{b}$ over $E\bar{c}\bar{a}$. As it was independent from $\bar{a}$ over $E\bar{c}\bar{b}$ by definition, we get that $\bar{a}'$ is independent from $(\bar{a}\bar{b})$ over $\text{acl}_\sigma(E\bar{c}\bar{a}) \cap \text{acl}_\sigma(E\bar{c}\bar{b})$ (by Corollary 4.22). Let $\bar{d} \in \text{acl}_\sigma(E\bar{c}\bar{a}) \cap \text{acl}_\sigma(E\bar{c}\bar{b})$ be such that $\bar{a}'$ and $\bar{a}\bar{b}$ are independent over $E\bar{c}\bar{d}$. By the claim, we know that $\bar{d} \notin \text{acl}_\sigma(E\bar{c})$, and this implies that

$$\text{SU}(\bar{a}/E\bar{c}\bar{d}) < n,$$

as $\bar{d} \in \text{acl}_\sigma(E\bar{c}\bar{a})$. We now obtain

$$n - 1 = \text{SU}(\bar{a}/E\bar{c}\bar{b}) \leq \text{SU}(\bar{a}/E\bar{c}\bar{d}) < \text{SU}(\bar{a}/E\bar{c}) = n,$$

so that $\text{SU}(\bar{a}/E\bar{c}\bar{d}) = n - 1$, which implies that $\text{SU}(\bar{d}/E\bar{c}) = 1$.  ⊣

**6.10. Definitions.** Let $S$ be an $(\infty-)$ definable set, defined over some $E = \text{acl}_\sigma(E)$. We say that $S$ is *modular* if, for every $m$ and $n$, and $\bar{a} \in S^m$, $\bar{b} \in S^n$, we have that $\bar{a}$ and $\bar{b}$ are independent over $\text{acl}_\sigma(E\bar{a}) \cap \text{acl}_\sigma(E\bar{b})$.

A type (over some set $E$) is *modular* if the set of its realizations is modular.

REMARKS. (1) In general, one requires that $\bar{a}$ and $\bar{b}$ are independent over $\mathrm{acl}^{\mathrm{eq}}(E\bar{a}) \cap \mathrm{acl}^{\mathrm{eq}}(E\bar{b})$. We use the fact that $\mathrm{Th}(K)$ eliminates imaginaries.

(2) This notion is also sometimes referred to as "1-basedness" in the case of stable theories.

(3) There is no harm in extending the notion to sets which are invariant under $E$-automorphism of $K$, i.e., sets of realizations of a set of types over $E$.

**6.11. Canonical bases.** Let $E = \mathrm{cl}_\sigma(E)$, $\bar{a}$ a tuple. Since $I_\sigma(\bar{a}/E)$ is finitely generated as a $\sigma$-ideal, there is an integer $m$ such that the (algebraic) locus $V$ of $(\bar{a}, \ldots, \sigma^m(\bar{a}))$ over $E$ completely describes $\mathrm{qftp}(\bar{a}/E)$.

That is, if $\bar{b}$ is such that $(\bar{b}, \ldots, \sigma^m(\bar{b}))$ has locus $V$ over $E$, then $\mathrm{qftp}(\bar{b}/E) = \mathrm{qftp}(\bar{a}/E)$.

Let $k_0$ be the field of definition of $V$, and let $k$ be the difference field generated by $k_0$. Then $\mathrm{tp}(\bar{a}/E)$ does not fork over $k$, and $k$ is in a sense smallest with that property (not quite true in positive characteristic). We call $k$ the *canonical base* of $\mathrm{tp}(\bar{a}/E)$ and denote it by $\mathrm{Cb}(\bar{a}/E)$.

**6.12. Comments.** (1) It is well known that if $\bar{c}_i$, $i \in \omega$ is an independent sequence of generics of $V$ over $E$, then $k_0$ is contained in the field generated by $\bar{c}_0, \ldots, \bar{c}_N$ for some $N$. Hence, the same is true in our case *provided $E$ is algebraically closed*: if $\bar{a}_i$, $i \in \omega$ is a sequence of independent realizations of $\mathrm{tp}(\bar{a}/E)$, then $k$ is contained in $\mathrm{cl}_\sigma(\bar{a}_0, \ldots, \bar{a}_N)$ for some $N$.

(2) Our definition does not quite agree with the usual definition of canonical bases for types in simple theories.

(3) What we are really interested in, is that $k^{\mathrm{alg}}$ is the smallest algebraically closed difference field over which $\mathrm{tp}(\bar{a}/E)$ does not fork. We could also have defined $\mathrm{Cb}(\bar{a}/E)$ as $k^{\mathrm{alg}}$. What really matters is that by (1), there is an integer $N$ such that if $\bar{a}_1, \ldots, \bar{a}_N$ are independent realizations of $\mathrm{tp}(\bar{a}/E)$ then $\mathrm{Cb}(\bar{a}/E) \subset \mathrm{acl}_\sigma(\bar{a}_1, \ldots, \bar{a}_N)$. See the remark in Subsection 4.23 for a proof.

**6.13. Corollary.** Let $S$ be modular. Then for every $n$, $m$, and $\bar{a} \in S^n$, $\bar{b} \in K^m$, we have that $\bar{a}$ and $\bar{b}$ are independent over $C = \mathrm{acl}_\sigma(E\bar{a}) \cap \mathrm{acl}_\sigma(E\bar{b})$.

PROOF. Let $\bar{a}_i$, $i \in \omega$ be a sequence of realizations of $\mathrm{tp}(\bar{a}/\mathrm{acl}_\sigma(E\bar{b}))$, independent over $E\bar{b}$, and with $\bar{a}_0 = \bar{a}$. Then $C = \mathrm{acl}_\sigma(E\bar{a}_i) \cap \mathrm{acl}_\sigma(E\bar{b})$ for $i \in \omega$ because $\mathrm{tp}(\bar{a}_i/E\bar{b}) = \mathrm{tp}(\bar{a}/E\bar{b})$. Since $\bar{a}$ is independent from $\{\bar{a}_i \mid i > 0\}$ over $E\bar{b}$, we have

$$C = \mathrm{acl}_\sigma(E\bar{a}) \cap \mathrm{acl}_\sigma(E\bar{a}_i \mid i > 0).$$

By the above, $\mathrm{acl}_\sigma(E\bar{a}_i \mid i > 0)$ contains the canonical base of $\mathrm{tp}(\bar{a}/\mathrm{acl}_\sigma(E\bar{b}))$. By modularity, we get that this canonical base is contained in $C$, and therefore that $\bar{a}$ and $\bar{b}$ are independent over $C$. ⊣

**6.14. Remark.** This is a particularly nice way of phrasing modularity: a set $S$ is modular if and only if, for every tuple $\bar{a}$ of elements of $S$ and set $B$, we have that $\mathrm{acl}_\sigma(E\bar a)\cap\mathrm{acl}_\sigma(EB)$ contains the canonical base of $\mathrm{tp}(\bar a/\mathrm{acl}_\sigma(EB))$. It coincides with the definition of 1-basedness.

**6.15. What is modularity?** Or rather, what is it not? Modularity forbids the existence of complicated sets. For instance, in the theory of algebraically closed fields, no infinite set is modular. This comes from the typical counterexample to modularity: let $a, b, c$ be algebraically independent over $\mathbb{Q}$, say, and let $d = ac + b$. Then

$$\mathbb{Q}(a,b)^{\mathrm{alg}} \cap \mathbb{Q}(c,d)^{\mathrm{alg}} = \mathbb{Q}^{\mathrm{alg}},$$

but $(a, b)$ and $(c, d)$ are not independent.

In case the elements of $S$ have SU-rank 1, modularity is equivalent to the nonexistence of a SU-rank-2 family of "curves", that is, a set $C(e)$, $e \in D$, of definable subsets of $S^2$ of SU-rank 1 where $D$ is definable, has SU-rank 2, and is such that if $d \neq d' \in D$, then $C(d) \cap C(d')$ is finite.

When we are in a stable situation and there is a group around, there is a remarkable theorem of Hrushovski-Pillay [14] which says:

THEOREM. Let $G$ be a stable group (maybe with additional structure) and assume that $G$ is modular. Then for every $n$, any definable subset of $G^n$ is a Boolean combination of cosets of definable subgroups of $G^n$, and these subgroups are defined over $\mathrm{acl}(\emptyset)$.

Thus in particular, a stable modular group has essentially only one group law. Moreover, one can show that it is necessarily abelian by finite.

The completions of ACFA are unstable, and we will not be able to apply directly the result of Hrushovski-Pillay. In fact, in positive characteristic this result will be *false*. However, all the quantifier-free formulas are "stable" and they control independence: (I will not define what a stable formula is; let me just say that if $T$ is a stable theory then all formulas are stable.) It will follow that some restricted version of Hrushovski-Pillay's theorem [11] holds in positive characteristic, while the full result holds in characteristic 0.

**6.16. Lemma.** Let $T$ be a complete theory satisfying the conclusion of Subsection 4.20(2), that if $A \underset{C}{\downarrow} B$ and $A \underset{B}{\downarrow} C$ then $A \underset{\mathrm{acl}(B)\cap\mathrm{acl}(C)}{\downarrow} BC$ (e.g., eliminating imaginaries, and stable or with stable forking), let $A, B, C$ be algebraically closed sets containing $E = \mathrm{acl}(E)$, and assume that $C$ is independent from $(A, B)$ over $E$. Then $\mathrm{acl}(AC) \cap \mathrm{acl}(BC) = \mathrm{acl}((A \cap B)C)$.

PROOF. Let $\alpha \in \mathrm{acl}(AC) \cap \mathrm{acl}(BC)$. By hypothesis, $C \underset{E}{\downarrow} AB$, so that $C \underset{A}{\downarrow} B$. Since $\alpha \in \mathrm{acl}(AC)$, we get $(C, \alpha) \underset{A}{\downarrow} B$. Similarly, $(C, \alpha) \underset{B}{\downarrow} A$. By the facts in Subsection 4.20, we obtain $(C, \alpha) \underset{A\cap B}{\downarrow}(A, B)$, which gives $\alpha \in \mathrm{acl}((A \cap B)C)$.                    ⊣

**6.17. Proposition.** Let $E = \mathrm{acl}_\sigma(E) \subset F = \mathrm{acl}_\sigma(F) \subset K$, let $\bar a$ be a tuple in $K$ which is independent from $F$ over $E$, and let $S$ be a set of realizations of a set of types of SU-rank 1 over $E$.

(1) If $\mathrm{tp}(\bar{a}/E)$ is modular, then so is $\mathrm{tp}(\bar{a}/F)$.

(2) $S$ is modular if and only if all types realized in $S$ are modular.

(3) If $\mathrm{SU}(\bar{a}/E) = 1$ and $\mathrm{tp}(\bar{a}/F)$ is modular, then $\mathrm{tp}(\bar{a}/E)$ is modular.

PROOF. (1) Let $\bar{b}$ be a finite set of realizations of $\mathrm{tp}(\bar{a}/F)$, and let $\bar{c}$ be a finite tuple of elements of $K$. By modularity, $\bar{b}$ is independent from $\mathrm{acl}_\sigma(F\bar{c})$ over $D = \mathrm{acl}_\sigma(E\bar{b}) \cap \mathrm{acl}_\sigma(F\bar{c})$. This implies that $\bar{b}$ is independent from $\mathrm{acl}_\sigma(F\bar{c})$ over $\mathrm{acl}_\sigma(FD) \subseteq \mathrm{acl}_\sigma(F\bar{b}) \cap \mathrm{acl}_\sigma(F\bar{c})$ and shows that $\mathrm{tp}(\bar{a}/F)$ is modular.

(2) One direction is clear: if $S$ is modular and $P \subset S$, then $P$ is modular. For the other direction, assume that this is not true, and let $n$ be minimal such that there are $\bar{a}_1, \ldots, \bar{a}_n \in S$, a tuple $\bar{b}$ in $K$ and $C = \mathrm{acl}_\sigma(E, \bar{a}_1, \ldots, \bar{a}_n) \cap \mathrm{acl}_\sigma(E\bar{b})$, with $(\bar{a}_1, \ldots, a_n)$ and $\bar{b}$ not independent over $C$.

Note that, by minimality of $n$ and because all tuples have SU-rank 1, we have that the elements $\bar{a}_1, \ldots, \bar{a}_n$ are independent over $E$. Moreover, the types $\mathrm{tp}(\bar{a}_i/E)$ are pairwise nonorthogonal: indeed, assume by way of contradiction that $\mathrm{tp}(\bar{a}_n/E)$ is orthogonal to $\mathrm{tp}(\bar{a}_1/E)$. By minimality of $n$ and transitivity of forking, we have that $(\bar{a}_1, \ldots, \bar{a}_{n-1})$ and $\bar{b}$ are independent over $C' = \mathrm{acl}_\sigma(E, \bar{a}_1, \ldots, \bar{a}_{n-1}) \cap \mathrm{acl}_\sigma(E\bar{b})$, and that $\bar{a}_n$ and $\bar{b}$ are not independent over $(C', \bar{a}_1, \ldots, \bar{a}_{n-1}) = (E, \bar{a}_1, \ldots, \bar{a}_{n-1})$. Hence $C = C'$, and $\bar{a}_n \in \mathrm{acl}_\sigma(E, \bar{a}_1, \ldots, \bar{a}_{n-1}, \bar{b}) \backslash \mathrm{acl}_\sigma(E, \bar{a}_1, \ldots, \bar{a}_{n-1})$. Since $\mathrm{tp}(\bar{a}_n/E)$ is orthogonal to $\mathrm{tp}(\bar{a}_1/E)$, we get that $\bar{a}_n \in \mathrm{acl}_\sigma(E, \bar{a}_2, \ldots, \bar{a}_{n-1}, \bar{b})$ which contradicts the minimality of $n$.

By definition and because $\mathrm{SU}(\bar{a}_i/E) = 1$, this means that for each $i \geq 2$ there is $F_i$ containing $E$ and independent from $\bar{a}_i$ over $E$ such that $\mathrm{acl}_\sigma(F_i\bar{a}_i) \backslash F_i$ contains a realization $\bar{a}_i'$ of $\mathrm{tp}(\bar{a}_1/E)$. Moving the $F_i$s by an $E$-automorphism, we may choose them such that $(F_2, \ldots, F_m)$ is independent from $(\bar{a}_1, \ldots, \bar{a}_n, \bar{b})$ over $E$. Setting $F = \mathrm{acl}_\sigma(F_2, \ldots, F_m)$ and using Lemma 6.16, we then have $\mathrm{acl}_\sigma(F, \bar{a}_1, \ldots, \bar{a}_n) \cap \mathrm{acl}_\sigma(F\bar{b}) = \mathrm{acl}_\sigma(FC)$, but $(\bar{a}_1, \ldots, \bar{a}_n)$ is not independent from $\bar{b}$ over $\mathrm{acl}_\sigma(FC)$.

As each $\bar{a}_i$ is equialgebraic over $F$ with the realization $\bar{a}_i'$ of $\mathrm{tp}(\bar{a}_1/E)$, we get that

— $\mathrm{acl}_\sigma(F, \bar{a}_1, \bar{a}_2', \ldots, \bar{a}_n') = \mathrm{acl}_\sigma(F, \bar{a}_1, \ldots, \bar{a}_n)$;

— $\mathrm{SU}(\bar{a}_1, \ldots, \bar{a}_n/F) = n = \mathrm{SU}(\bar{a}_1, \bar{a}_2', \ldots, \bar{a}_n'/F) = \mathrm{SU}(\bar{a}_1, \bar{a}_2', \ldots, \bar{a}_n'/E)$.

Therefore (by transitivity of independence), we get that

$$F \text{ and } (\bar{a}_1, \bar{a}_2', \ldots, \bar{a}_n', \bar{b}) \text{ are independent over } E. \qquad (*)$$

Hence, letting $D = \mathrm{acl}_\sigma(E, \bar{a}_1, \bar{a}_2', \ldots, \bar{a}_n') \cap \mathrm{acl}_\sigma(E\bar{b})$, by Lemma 6.16 we have

$$\mathrm{acl}_\sigma(F, a_2', \ldots, \bar{a}_n') \cap \mathrm{acl}_\sigma(F\bar{b}) = \mathrm{acl}_\sigma(FD) = \mathrm{acl}_\sigma(FC).$$

By modularity of $\mathrm{tp}(\bar{a}_1/E)$, we also have that $(\bar{a}_1, \bar{a}_2', \ldots, \bar{a}_n')$ and $\bar{b}$ are independent over $D$, and by $(*)$, this gives that they are independent over $\mathrm{acl}_\sigma(FD)$. But, since each $\bar{a}_i'$ is equialgebraic with $\bar{a}_i$ over $F$, we have that $\bar{a}_1, \bar{a}_2', \ldots, \bar{a}_n'$

and $\bar{b}$ are not independent over $\mathrm{acl}_\sigma(FC) = \mathrm{acl}_\sigma(FD)$: this gives us the desired contradiction.

(3) Assume that $\mathrm{tp}(\bar{a}/F)$ is modular. Note that if $\varphi$ is an $E$-automorphism of $K$, then $\varphi(\mathrm{tp}(\bar{a}/F))$ is also modular. Let $\bar{a}_1, \ldots, \bar{a}_n$ be realizations of $\mathrm{tp}(\bar{a}/E)$, $\bar{b}$ a tuple of elements of $K$, and $C = \mathrm{acl}_\sigma(E\bar{a}_1, \ldots, \bar{a}_n) \cap \mathrm{acl}_\sigma(E\bar{b})$. For each $i$, let $F_i$ be such that $\mathrm{tp}(\bar{a}_i, F_i/E) = \mathrm{tp}(\bar{a}, F/E)$; then each $\mathrm{tp}(\bar{a}_i/F_i)$ is modular. Using (1) and moving the $F_i$ by some $E$-automorphism of $K$, there is $F'$ independent from $(\bar{a}_1, \ldots, \bar{a}_n, \bar{b})$ over $E$ such that $\mathrm{tp}(\bar{a}_i/F')$ is modular for $i = 1, \ldots, n$. By Lemma 6.16, we have $\mathrm{acl}_\sigma(F'\bar{a}_1, \ldots, \bar{a}_n) \cap \mathrm{acl}_\sigma(F'\bar{b}) = \mathrm{acl}_\sigma(F'C)$. By (2), we get that $(\bar{a}_1, \ldots, \bar{a}_n)$ and $\bar{b}$ are independent over $\mathrm{acl}_\sigma(F'C)$. Since $F'$ is independent from $(\bar{a}_1, \ldots, \bar{a}_n, \bar{b})$ over $E$, this implies that $(\bar{a}_1, \ldots, \bar{a}_n)$ is independent from $\bar{b}$ over $C$ and therefore that $\mathrm{tp}(\bar{a}/E)$ is modular.     ⊣

**6.18. Corollary.** Let $p$ and $q$ be types over algebraically closed sets which are of SU-rank 1 and are nonorthogonal. Then $p$ is modular if and only if $q$ is modular.

PROOF. Let $p'$ and $q'$ be nonforking extensions of $p$ and $q$ to some set $A = \mathrm{acl}_\sigma(A)$ containing the sets over which $p$ and $q$ are defined, and such that $p'$ is non-almost-orthogonal to $q'$. Use Proposition 6.17.     ⊣

**6.19. Nonexample.** Let $E = \mathrm{acl}_\sigma(E) \subset K$, let $\bar{a}$ be a tuple, and assume that $\mathrm{SU}(\bar{a}/E) \geq \omega$. We claim that $\mathrm{tp}(\bar{a}/E)$ is not modular.

Indeed, we know that some element of the tuple $\bar{a}$, say $a_1$, is transformally transcendental over $E$. Take $\bar{b}$ realizing $\mathrm{tp}(\bar{a}/E)$ and independent from $\bar{a}$ over $E$, and let $c$ be transformally transcendental over $\mathrm{acl}_\sigma(E\bar{a}\bar{b})$, let $d = a_1 c + b_1$. Then certainly $(\bar{a}, \bar{b})$ and $(c, d)$ are not independent over $E$. Since $d = a_1 b + b_1$, we have that $(c, d)$ is independent from $(\bar{a}, \bar{b})$ over $\mathrm{acl}_\sigma(Ea_1b_1)$, and so it is enough to show that $(a_1, b_1)$ and $(c, d)$ are not independent over $C = \mathrm{acl}_\sigma(Ea_1b_1) \cap \mathrm{acl}_\sigma(Ecd)$.

We claim that the canonical base of $\mathrm{tp}(c, d/Ea_1b_1)$ contains $a_1b_1$. Indeed, let $(c', d')$ be a realization of $\mathrm{tp}(c, d/Ea_1b_1)$ independent from $(a_1, b_1, c, d)$ over $Ea_1b_1$. Then $a_1, b_1, c, c'$ are independent over $E$. However, $(a_1, b_1)$ belongs to the field generated by $(c, d, c_1, d_1)$. As $(a_1, b_1)$ does not belong to $\mathrm{acl}_\sigma(Ecd)$, this implies that $(a_1, b_1)$ and $(c, d)$ are not independent over $C = \mathrm{acl}_\sigma(Ea_1b_1) \cap \mathrm{acl}_\sigma(Ecd)$.

**6.20. Sets of SU-rank 1.** Let us again concentrate on sets $S$ of SU-rank 1 defined over $E$. Because all our tuples have SU-rank 1, we get a pregeometry on $S$ where the closure operator is simply $\mathrm{acl}(E, -)$: given $A \subset S$ and $b \in S$, either $b \in \mathrm{acl}(E, A)$ or $b$ is independent from $A$ over $E$. The exchange principle holds, and the SU-rank of a set is the dimension of this set.

DEFINITION. An SU-rank-1 type $p$ over $E$ is called *trivial* if whenever $\bar{a}_1, \ldots, \bar{a}_n, \bar{a}$ realize $p$, then either $\bar{a} \notin \mathrm{acl}_\sigma(E, \bar{a}_1, \ldots, \bar{a}_n)$, or $\bar{a} \in \mathrm{acl}_\sigma(E\bar{a}_i)$ for some $i$.

**6.21. Exercise 14.** Show that a trivial type is modular.

### 6.22. Some additional properties of modular types in models of ACFA.

(1) If the type $p$ is nonorthogonal to a modular type of SU-rank 1, then $p$ is non-almost-orthogonal to a (modular) type of SU-rank 1.

(2) If $p$ is a nontrivial modular type of SU-rank 1, then $p$ is nonorthogonal to the generic of a group of SU-rank 1 defined over the same set as $p$ (see the definition of generic in the next section).

**§7. Groups, generic types, stabilizers.** In this section, contrary to my promise, I will use some results on simple theories. You may very well replace everywhere "simple", or "supersimple", by complete theory extending ACFA. The notion of generic is meant to be the analogue of a generic of an algebraic group (so generic in the sense defined in Subsection 2.10). A good reference for generics and stabilizers in simple theories is Pillay [16]. But see also Wagner [19].

**7.1. Generics in groups.** Let $G$ be a group (maybe with extra structure) whose theory is simple (or let $G$ be a group definable in some model of ACFA). We call a type $p$ over some set $E = \mathrm{acl}(E)$ a *generic type of $G$* if $p$ is realized in $G$, and whenever $a \in G$ and $b$ realizing $p$ are independent over $E$, then $b \cdot a$ is independent from $a$ over $E$. An element of $G$ is *generic over $E$* if it realizes a generic type over $E$.

Let $H$ be a definable subgroup of $G$, let $a \in G$, consider the coset $a \cdot H$ and let $E = \mathrm{acl}(E)$ be a set over which $a \cdot H$ is defined. We say that $b$ is a *generic of the coset $a \cdot H$ over $E$* if $b \in a \cdot H$ and whenever $h \in H$ is independent from $b$ over $E$, then $b \cdot h$ is independent from $h$ over $E$.

SOME FACTS.

(1) Generics exist. A nonforking extension of a generic type is generic.

(2) Assume that $\mathrm{tp}(b/E)$ is generic, and let $a \in G$ be independent from $b$ over $E$. Then $a \cdot b$ is independent from $a$ over $E$, and $a \cdot b$ and $b \cdot a$ realize generic types over $E$. (The definition described a generic element of $G$ in terms of its action on $G$ by multiplication on the left. This shows that it can be defined also in terms of the action by multiplication on the right, and that a "left" generic is also a "right" generic.)

Below we will give some examples illustrating this notion. First we need some definitions from algebraic geometry.

**7.2. Projective space, projective varieties.** Let $n$ be an integer, and consider the set of lines through the origin $O$ in affine $n + 1$-space $\mathbb{A}^{n+1}(\Omega)$. This set can be viewed as the set of equivalence classes of $\mathbb{A}^{n+1}(\Omega) \setminus O$ for the equivalence relation

$$(x_0, \ldots, x_n) \sim (y_0, \ldots, y_n) \iff \exists \lambda \neq 0 \bigwedge_{i=0}^{n+1} x_i = \lambda y_i.$$

We denote this set by $\mathbb{P}^n(\Omega)$ and call it the projective $n$-space. The equivalence class of $(x_0, \ldots, x_n)$ is usually denoted by $(x_0 : \cdots : x_n)$.

If $F(x_0, \ldots, x_n)$ is a *homogeneous polynomial* over $\Omega$, i.e., for some $d$ all monomials appearing in $F$ have total degree equal to $d$, then the equation $F(x_0, \ldots, x_n) = 0$ is compatible with the equivalence relation and defines a subset of $\mathbb{P}^n(\Omega)$. The topology whose closed sets are finite intersections of sets of this form is called the Zariski topology.

Note that we can view $\mathbb{P}^n(\Omega)$ as the union of $n + 1$ copies of affine $n$-space. Indeed, for $i = 0, \ldots, n$, define

$$U_i = \{(x_0 : \cdots : x_n) \mid x_i \neq 0\}.$$

Then the map $f_i : (x_1, \ldots, x_n) \mapsto (x_1 : \cdots : x_{i-1} : 1 : x_i : \cdots : x_n)$ defines a bijection between $\mathbb{A}^n(\Omega)$ and $U_i$. Moreover, $\mathbb{P}^n(\Omega) = \bigcup_{i=1}^n U_i$. Each $U_i$ is open for the Zariski topology, and the maps $f_i$ are homeomorphisms (for the Zariski topology on $\mathbb{A}^n(\Omega)$ and the induced topology on $U_i$).

Closed subsets of projective $n$-space are called projective sets. If $X$ is closed and irreducible (i.e., for each $i$, $U_i \cap X$ is a variety), then we call $X$ a projective variety. One can show that the product of two projective varieties is a projective variety.

Morphisms between projective varieties are defined locally using the covering by affine spaces.

**7.3. Algebraic groups.** We say that $G$ is an *algebraic group* if $G$ is an open subset of a projective variety, and we have a group operation $G \times G \to G$ which is a morphism, i.e., is everywhere defined, and is locally defined by rational functions. We also require that the inverse map $G \to G$ is a morphism.

Here are some examples:

— The additive group, usually denoted $\mathbb{G}_a$ to distinguish it from the affine line $\mathbb{A}^1$ (with no structure). It is an open subset of $\mathbb{P}^1$.

— The multiplicative group $\mathbb{G}_m$ (so $\mathbb{G}_m(K) = K \setminus \{0\}$).

— Also, some projective group varieties (i.e., closed for the Zariski topology) which are called *abelian varieties*. Some examples are elliptic curves, and also certain groups of the form $\mathbb{C}^n/\Lambda$ where $\Lambda$ is a $2n$-dimensional lattice subgroup of $\mathbb{C}^n$ generating the $\mathbb{R}$-vector space $\mathbb{C}^n$.

**7.4. Some elementary facts about algebraic groups.** We did not require in our definition that $G$ be irreducible (by which I mean that the Zariski closure of $G$ in the projective space we are working in is a variety).

If $G$ is not irreducible, then one can verify that the irreducible components of $G$ are disjoint so that there is a unique irreducible component of $G$, denoted $G^0$ and called the *connected component of* $G$, which contains the identity element $e$ of $G$. Then $G^0$ is a subgroup of finite index of $G$.

Let $G$ be an algebraic group, and $H$ a subgroup of $G$, $\bar{H}$ its Zariski closure. Then $\bar{H}$ is also an algebraic group.

**7.5. Examples of generics.** We let $K$ be an algebraically closed field, sufficiently saturated, and $G$ an algebraic group defined over some subfield $E$ of $K$.

The first thing to remark is that a generic of $G(K)$ for the theory ACF of algebraically closed fields is simply a generic in the sense of algebraic geometry. This is fairly immediate from the definition: let $d = \dim(G)$, assume that $g$ is a generic of $G$, independent from $h \in G$ over $E$. We know that $g \cdot h \in E(g, h)$ and (using the inverse map) that $g \in E(g \cdot h, h^{-1}) = E(g \cdot h, h)$. Hence $g$ and $g \cdot h$ are equialgebraic over $E(h)$. This implies

$$\text{tr.deg}(g \cdot h/E(h)) = \text{tr.deg}(g/E(h)) = d \leq \text{tr.deg}(g \cdot h/E) \leq d = \dim(G),$$

so that equality holds and $g \cdot h$ is independent from $h$ over $E$, as desired.

Let us now assume that $K$ is a model of ACFA, sufficiently saturated, and let $H$ be a definable subgroup of $G(K)$, defined also over $E = \text{acl}_\sigma(E)$. Then the $\sigma$-closure $\tilde{H}$ of $H$ is a $\sigma$-closed subgroup of $G(K)$. Thus there is a natural notion of generic for $H$: an element $g \in H$ is generic if and only if $I_\sigma(g/E) = I_\sigma(\tilde{H}/E)$. Using the description of generics of algebraic groups, one can show (easily if $\text{SU}(H) < \omega$, with a little more work in the general case) that this definition is the correct one. Furthermore, that $[\tilde{H} : H] < \infty$ so that a generic of $H$ is also a generic of $\tilde{H}$.

One interesting consequence of elimination of imaginaries is the following: let $\tilde{H}^0$ be the connected component of $\tilde{H}$ (i.e., the irreducible component in the sense of the $\sigma$-topology which contains the identity element). Then both $\tilde{H}$ and $\tilde{H}^0$ are defined over $E$. The cosets of $\tilde{H}^0$ in $\tilde{H}$ are then imaginary elements and algebraic over $E$ (since $[\tilde{H} : \tilde{H}^0] < \infty$). By elimination of imaginaries, they are defined over $E$.

**7.6. Lemma.** Let $G$ be a group, $H$ a definable subgroup of $G$ defined over some $E = \text{acl}(E)$, and $g \in G$ a generic of $G$ over $E$. Then $g$ is a generic of $g \cdot H$ (over $C = E \cup \bar{c}$, where $\bar{c}$ is the code of $g \cdot H$).

PROOF. Let $h$ be a generic of $H$ over $Eg$. We claim that $g' = g \cdot h$ is a generic of $g \cdot H$ over $Eg$. Indeed, let $h_1 \in H$ be independent from $g'$ over $Eg$. Then $g' \cdot h_1 \in g \cdot H$, and we need to show that $h_1$ and $g' \cdot h_1$ are independent over $Eg$. But, $g' \cdot h_1$ is equialgebraic with $h \cdot h_1$ over $Eg$, and it is therefore enough to show that $h_1$ and $h \cdot h_1$ are independent over $Eg$, which follows because $h$ is a generic of $H$ over $Eg$. As $C \subset \text{acl}(Eg)$, this implies that $g'$ is a generic of $g \cdot H$ over $C$.

By genericity of $g$, we know that $g'$ and $h$ are independent over $E$ so that $h$ is also a generic of $H$ over $Eg'$. From $g \cdot H = g' \cdot H$ and the previous step, we deduce that $g$ is a generic of $g\dot{H}$ over $C$. ⊣

**7.7. Stabilizers.** Let $G$ be a group (maybe with extra structure) whose theory is simple, and let $p$ be a type over $E = \text{acl}(E)$. Assume that the independence theorem is satisfied over $E$; i.e., given $a$ and $b$ independent over $E$, and any two nonforking extensions $p_1$ to $(E, a)$ and $p_2$ to $(E, b)$ of some type $p$ over $E$, $p_1 \cup p_2$ extend to some type $p'$ over $(E, a, b)$ which does not fork over $E$ (or let $G$ be a group definable in a model of ACFA).

We define $S(p)$ to be the set of elements $h \in G$, such that *there exists* a realization $g$ of $p$ such that $h \cdot g$ realizes $p$, and $h$ is independent from $g$ and from $h \cdot g$ over $E$.

One then defines $\mathrm{Stab}(p) = S(p) \cdot S(p)$. By the independence theorem over $E$, we have that if $h_1$ and $h_2$ are independent elements of $S(p)$, then $h_1 \cdot h_2 \in S(p)$. This will imply that $\mathrm{Stab}(p)$ is a subgroup of $G$ with generics in $S(p)$. One can show that $\mathrm{Stab}(p)$ is $\infty$-definable.

**7.8. Example.** Let $K$ be a model of ACFA, $G$ an algebraic group defined over $E = \mathrm{acl}_\sigma(E)$, and $p$ be a type over $E$, $X \subseteq G(K)$ its set of realizations. By definition, if $h \in S(p)$, then there are $g, g_1 \in X$ such that $g \downarrow_E h$, $g_1 \downarrow_E h$, and $h \cdot g = g_1$. In particular, $h = g_1 \cdot g^{-1}$, so that $h \in X \cdot X^{-1}$. The requirement that $h$ be independent from $g$ and from $g_1$ over $E$ then implies that necessarily $\mathrm{SU}(h/E) \leq \mathrm{SU}(g/E) = \mathrm{SU}(g_1/E)$.

Let $Y$ be the $\sigma$-closure of $X$. Then $Y$ is irreducible, and $g$, $g_1$ belong to $Y$. This implies in particular that $S(p) \subset \mathrm{Stab}(Y) =_{\mathrm{def}} \{h \in G(K) \mid hY = Y\}$. Note that $\mathrm{Stab}(Y)$ is clearly a subgroup of $G(K)$ and is quantifier-free definable so that it is $\sigma$-closed. If I am not mistaken, one can then show that $\mathrm{Stab}(p)$ is the intersection of all definable subgroups of $\mathrm{Stab}(Y)$ which are defined over $E$ and have finite index in $\mathrm{Stab}(Y)$.

**7.9. Remarks/Exercise 15.** Let us now assume that $p$ has a unique non-forking extension to any set containing $E$, and let $P$ be the set of realizations of $p$. (This hypothesis is verified for instance when $p$ is definable over $E$ and $T$ is stable.)

(1) Then $\mathrm{Stab}(p) = S(p)$: this is because the quantifier "there exists" in the definition of $S(p)$ can be replaced by the quantifier "for all".

(2) If $E' \supset E$, and $p'$ is the nonforking extension of $p$ to $E'$, then $\mathrm{Stab}(p) = \mathrm{Stab}(p')$.

(3) Let $h \in E$. Then $h \in S(p) \iff h \cdot P = P$.

No completion of ACFA is stable. However, in the language of "local stability theory", all quantifier-free formulas are stable. We will see below some of the consequences this has.

**7.10. Proposition.** Let $E = \mathrm{acl}_\sigma(E) \subset K$, $K$ a sufficiently saturated model of ACFA.

(1) Let $p$ be a type over $E$, and assume that whenever $F$ contains $E$, then $p$ has a unique nonforking extension to $F$. Then $p$ is definable over $E$, i.e., given a formula $\varphi(\bar{x}, \bar{y})$, there is a formula (over $E$) $d_\varphi(\bar{y})$ which defines in $K$ the set of $\bar{c}$ such that $\varphi(\bar{x}, \bar{c})$ belongs to the unique-nonforking extension of $p$ to $K$.

(2) Let $p$ be a quantifier-free type over $E$. Given a quantifier-free formula $\varphi(\bar{x}, \bar{y})$, there is a quantifier-free formula (over $E$) $d_\varphi(\bar{y})$ which defines in $K$ the set of $\bar{c}$ such that $\varphi(\bar{x}, \bar{c})$ belongs to some/all nonforking extensions of $p$ to $K$.

PROOF. (1) is well known but we will give the proof. Assume by way of contradiction that $p$ is not definable, i.e., that there is some formula $\varphi(\bar{x}, \bar{y})$ such that the set $X$ of tuples $\bar{c}$ of $K$ such that $\varphi(\bar{x}, \bar{c})$ belongs to the unique nonforking extension of $p$ to $K$ is not definable. Then, for every $\mathcal{L}(E)$-formula $\psi(\bar{y})$, there are $\bar{c}_1 \in X$, $\bar{c}_2 \notin X$ which both satisfy $\psi(\bar{y})$. By compactness, there are $\bar{c}_1, \bar{c}_2$ in $K$, realizing the same type over $E$, and such that $\bar{c}_1 \in X$, $\bar{c}_2 \notin X$. Let $f$ be an automorphism of $K$ which leaves $E$ fixed and sends $\bar{c}_1$ to $\bar{c}_2$. If $q$ is the unique nonforking extension of $p$ to $K$, then $f(q)$ is also a nonforking extension of $p$ to $K$ and so must equal $q$: but this contradicts the fact that $\varphi(\bar{x}, \bar{c}_1) \in q$, $\varphi(\bar{x}, \bar{c}_2) \notin q = f(q)$.

We now sketch a proof of (2). The crucial point is the following observation: since quantifier-free types correspond exactly to descriptions of the isomorphism type of the "difference field generated by", it follows that if $F = \text{acl}_\sigma(F)$ contains $E$, then there is a unique quantifier-free type $q$ over $F$ which extends $p$ and is such that whenever $\bar{a}$ realizes $q$, then $\bar{a}$ is independent from $F$ over $E$.

This implies that, given $p = \text{qftp}(\bar{a}/E)$ and $\bar{c}$ independent from $\bar{a}$ over $E$, we have

$$\text{qftp}(\bar{a}/E) \cup \text{qftp}(\bar{c}/E) \cup \Sigma(\bar{x}, \bar{y}) \vdash \text{qftp}(\bar{a}, \bar{c}/E)$$

where $\Sigma(\bar{x}, \bar{y})$ is the set of quantifier-free formulas over $E$ expressing that the tuples $\bar{x}$ and $\bar{y}$ are independent over $E$. The proposition follows by compactness.                                                                    ⊣

**7.11. Remarks.** (1) Let $E = \text{acl}_\sigma(E)$, $\bar{a}$ a tuple in $K$. Let $C$ be the set of codes of all (quantifier-free) definitions of $\text{qftp}(\bar{a}/E)$. Then certainly $\text{tp}(\bar{a}/E)$ does not fork over $C$. Moreover, if $D \subset E$ and $\text{tp}(\bar{a}/E)$ does not fork over $D$ then $\text{acl}_\sigma(D) \supset C$. Hence, we are getting that for quantifier-free types, our definition of a canonical base agrees with the classical one.

(2) Let $G$ be a group definable in $K$ by quantifier-free formulas (over some $E = \text{acl}_\sigma(E)$), and assume in addition that the group operation and the inverse map are piecewise definable by terms of the language $\{+, -, \cdot, {}^{-1}, \sigma, \sigma^{-1}, e (e \in E)\}$ so that if $a, b \in G$, then $a \cdot b \in \text{cl}_\sigma(E, a, b)$ and $a^{-1} \in \text{cl}_\sigma(E, a)$. (By the multiplication being piecewise defined, we mean that there is a definable finite partition of $G^2$, and that on each piece of this partition multiplication is defined by a term.) If $p$ is a quantifier-free type over $E$ realized in $G$, then we may also define $\text{Stab}(p)$ to be the set of elements $b$ of $G$ such that whenever $a$ realizes $p$ and is independent from $b$ over $E$, then $b \cdot a$ realizes $p$ and is independent from $b$ over $E$. All remarks made in Subsection 7.9 go through

to this particular case. Examples of such groups are algebraic groups defined over $E$.

**7.12. Theorem.** Let $G$ be a group, quantifier-free definable in a model $K$ of ACFA, and assume that $G$ is modular, and the group operation and the inverse map are piecewise defined by terms of the language $\{+, -, \cdot, {}^{-1}, \sigma, \sigma^{-1}, e(e \in E)\}$. If $X$ is a quantifier-free definable subset of $G$, then $X$ is a Boolean combination of quantifier-free definable subgroups of $G$. Furthermore, if $G$ is defined over $E = \mathrm{acl}_\sigma(E)$, then so are these subgroups.

PROOF. We will first show that if $p$ is a quantifier-free type defined over some $A = \mathrm{acl}_\sigma(A)$ containing $E$, then the set $P$ of realizations of $p$ is contained in a coset of some quantifier-free definable subgroup $S$ of $G$, and $p$ is the generic type of this coset. Without loss of generality, we will assume that $A$ is the algebraic closure (over $E$) of the canonical base of $p$.

Define $S = \mathrm{Stab}(p)$, the set of elements $b \in G$ such that there is $a \in P$, independent from $b$ over $A$ and such that $b \cdot a \in P$.                                    ⊣

CLAIM. $S$ is quantifier-free definable (over $A$).

PROOF OF CLAIM. Let $a \in P$ and consider $I_\sigma(a/A)$. Then $I_\sigma(a/A)$ completely determines $p$: let $b \in G$. Then

$$I_\sigma(b/A) = I_\sigma(a/A) \iff b \in P.$$

Let $F(X)$ be a tuple of difference polynomials over $A$ generating the $\sigma$-ideal $I_\sigma(a/A)$ (as a $\sigma$-ideal).

By the piecewise quantifier-free definability of the group operation, there is a quantifier-free formula $\psi(y, x)$ such that, if $a \in P$ is independent from $g \in G$ over $A$, then

$$K \models \psi(g, a) \iff F(g \cdot a) = 0.$$

Let $\varphi$ be the formula $F(x) = 0 \wedge \psi(y, x)$. Then $S$ is defined by the formula $d_\varphi(y)$.

Let $g \in G$. Then $g$ satisfies $d_\varphi(y)$ if and only if $F(x) = 0 \wedge \psi(g, x)$ belongs to the nonforking extension of $p$ to $A \cup g$, if and only if whenever $a \in P$ is independent from $g$ over $A$, then $b = g \cdot a$ satisfies $F(x) = 0$. Note that because $b$ and $a$ are equialgebraic over $\mathrm{acl}_\sigma(A, g)$, this implies that $b$ belongs to $P$ and is independent from $g$ over $A$ (because $b$ cannot satisfy "more" equations than $F(x)$ over $\mathrm{acl}_\sigma(A, g)$).

(In the classical stable (nonsuperstable) case, one cannot get $S$ to be definable, only $\infty$-definable. There the defining formulas are of the form $\forall \bar{z}\, d_\varphi(\bar{z}) \leftrightarrow d_{\varphi^*}(\bar{z}, y)$, where $\varphi^*(\bar{z}, y, x) = \varphi(\bar{z}, y \cdot x)$.)

Fix $a \in P$, and $g$ a generic of $G$ over $Aa$. Then $b = g \cdot a$ is a generic of $G$ over $Aa$ and is independent from $a$ over $A$. Let $G_0 \prec G$ contain $A, g$, let $P_0$ be the set of realizations of the nonforking extension of $p$ to $G_0$, and $Q_0$ the set of realizations of the nonforking extension $q_0$ of $\mathrm{qftp}(b/Ag)$ to $G_0$. Then

$Q_0 = g \cdot P_0$. If $\tau \in \mathrm{Aut}(G_0/A)$, then $\tau$ also acts on the set of formulas over $G_0$. By abuse of notation, if $U$ is an $(\infty\text{-})$definable over $G_0$ subset of $G$, we will denote by $\tau(U)$ the $(\infty\text{-})$definable subset of $G$ defined by applying $\tau$ to the formulas defining $U$. ⊣

CLAIM. Let $\tau \in \mathrm{Aut}(G_0/A)$. Then $\tau(Q_0) = Q_0 \iff \tau(g \cdot S) = g \cdot S$.

PROOF OF CLAIM. Since $p$ is definable over $A$, we have that $\tau(P_0) = P_0$ and $\tau(S) = S$. Hence $\tau(Q_0) = Q_0 \iff \tau(g) \cdot P_0 = g \cdot P_0 \iff g^{-1} \cdot \tau(g) \cdot P_0 = P_0 \iff g^{-1} \cdot \tau(g) \in S \iff \tau(g \cdot S) = g \cdot S$. ⊣

By elimination of imaginaries of ACFA, we get that the codes of $Q_0$ and of $g \cdot S$ are equialgebraic over $A$. Let $\bar{c}$ be the code of $g \cdot S$. Because $g$ is a generic of $G$ over $A$, and $S$ is a subgroup of $G$ defined over $A$, we get that $g$ is a generic of $g \cdot S$ (see Lemma 7.6), and that $\bar{c}$ is equialgebraic with the canonical base of $\mathrm{qftp}(g/\mathrm{acl}_\sigma(A\bar{c}))$, so that $\mathrm{SU}(g/A\bar{c}) = \mathrm{SU}(S)$. By the claim, we also get that the canonical base of $q_0$ is equialgebraic over $A$ with $\bar{c}$. Since $q_0$ is the nonforking extension of $\mathrm{qftp}(b/\mathrm{acl}_\sigma(Ag))$ to $G_0$, it follows that $\bar{c}$ is equialgebraic over $A$ with $\mathrm{Cb}(b/Ag)$. By modularity, we have

$$\mathrm{acl}_\sigma(A\bar{c}) = \mathrm{acl}_\sigma(Ab) \cap \mathrm{acl}_\sigma(Ag). \qquad (**)$$

Since $g$ is a generic of $G$ over $Aa$, we have that $a$ is independent from $g$ and from $b = g \cdot a$ over $A$, and therefore also over $A\bar{c}$ by $(**)$. Hence, using the symmetry of independence, we have

$$\mathrm{SU}(P) = \mathrm{SU}(b/Ag) = \mathrm{SU}(b/A\bar{c}) = \mathrm{SU}(b/A\bar{c}a)$$
$$= \mathrm{SU}(g/A\bar{c}a) = \mathrm{SU}(g/A\bar{c}) = \mathrm{SU}(S).$$

(The fourth equality is because $b$ and $g$ are equialgebraic over $Aa \subset Aa\bar{c}$.) Hence $\mathrm{SU}(P) = \mathrm{SU}(S)$. If $h \in S$ is independent from $a$ over $A$, then $b = h \cdot a \in P$, and is independent from $a$ and from $h$ over $A$. Hence the generics of $S$ are of the form $d \cdot e^{-1}$, for $d, e \in P$ independent over $A$. This implies that $Sd = Sa$ for all $d \in P$, and therefore that $P \subset Sa$. It remains to show that if $b$ is a generic of $Sa$, then $b \in P$. Let $b$ be a generic of $Sa$ over $A$, and choose $d \in P$ independent from $b$ over $A$. Then $Sd = Sa$ so that there is $h \in S$ such that $b = h \cdot d$. From $\mathrm{SU}(S) = \mathrm{SU}(b/A) = \mathrm{SU}(b/Ad) = \mathrm{SU}(h/Ad) \leq \mathrm{SU}(S)$, we deduce that $h$ is independent from $d$ over $A$ so that $h \cdot d \in P$ (by definition of $S$).

Hence the type $p$ can be described as follows:

— $x$ belongs to the coset $Sa$;

— if $S'$ is a definable subgroup of $G$, and the coset $S'b$ is definable over $A$ and strictly contained in $Sa$, then $x$ does not belong to $S'b$.

Indeed, if $b \in Sa$ is *not* a generic of $Sa$, then the realizations of $\mathrm{qftp}(b/A)$ are the generics of some $S'b$ strictly contained in $Sa$ and defined over $A$. It follows, by compactness, that every formula is equivalent to a Boolean combination of formulas of the form "$x$ belongs to a coset of some definable subgroup of $G$".

We will now show that $S$ is in fact defined over $E$. By modularity, $\mathrm{tp}(g/A\bar{c})$ is definable over $\mathrm{acl}_\sigma(Eg)$, and since $g \cdot S$ is a coset of a group, this implies that $g \cdot S$ is also defined over $\mathrm{acl}_\sigma(Eg)$. Hence so is $S = g^{-1} \cdot (g \cdot S)$, and we obtain that $S$ is defined over $\mathrm{acl}_\sigma(Eg) \cap A = E$.                           ⊣

### §8. General results about models of ACFA.

We state here some results without proofs. We are working in a model $K$ of ACFA. Let us start with an easy one.

**8.1. Proposition.** Let $\phi$ be the identity if $\mathrm{char}(K) = 0$, and the Frobenius automorphism $x \mapsto x^p$ if $\mathrm{char}(K) = p > 0$. Let $m \geq 1$ and $n$ be integers. Then the difference field $(K, \sigma^m \phi^n)$ is also a model of ACFA.

PROOF. See below Exercise 8.19.                           ⊣

**8.2. The dichotomy theorem.** Using Theorem 5.1, we obtain that in positive characteristic the fields $\mathrm{Fix}(\sigma^m \phi^n)$ are pseudofinite fields. We will refer to these fields as *the fixed fields of $K$.*

Assume that $(m, n) = 1$ (or $m = 1, n = 0$) and let $F = \mathrm{Fix}(\sigma^m \phi^{-n})$. One can then show that $\mathrm{SU}(F) = 1$. From this, it follows that if $S \subset F$ is definable and infinite, then $F = a_1 S + a_2 S + \cdots + a_n S$ for some elements $a_1, \ldots, a_n \in K$. This implies that all nonalgebraic 1-types containing the formula $\sigma(x) = x$ (or $\sigma^m(x) = \phi^n(x)$ if the characteristic is positive) are nonorthogonal; since the difference field $F$ is certainly nonmodular and has no induced structure from $K$, this implies the following: A type $p$ which is non-orthogonal to one of the formulas $\sigma^m(x) = \phi^n(x)$ cannot be modular.

In fact, the converse is also true.

THEOREM. Let $K$ be a model of ACFA, and let $p$ be a type of SU-rank 1 over $E = \mathrm{acl}_\sigma(E) \subset K$. Then $p$ is modular if and only if $p$ is orthogonal to all formulas $\sigma^m(x) = \phi^n(x)$.

If $\mathrm{char}(K) = 0$, then $p$ is modular if and only if $p$ has a unique nonforking extension to any set containing $E$.

**8.3. Discussion/theorem** □□□. In characteristic 0, we therefore get that modular types of SU-rank 1 are definable and hence stable. Using Proposition 6.9, one can then show that if a (finite SU-rank) formula $\varphi(\bar{x})$ is orthogonal to $\sigma(x) = x$ (i.e., to all types containing the formula $\sigma(x) = x$), then all types containing $\varphi(\bar{x})$ are stable and definable, the set $S$ defined by $\varphi$ is *stably embedded*, i.e., all subsets of $S^m$ which are definable in $K$ are definable with parameters from $S$. Hence $S$ with the structure induced from $K$ is *superstable of finite U-rank and one-based.*

In characteristic $p > 0$, if a finite SU-rank formula is orthogonal to all formulas defining fixed fields, then the set it defines is also modular but is usually not stable.                                                     □□□

**8.4.** For $p$ a prime and $q$ a power of $p$, consider the difference field $F_q = (\mathbb{F}_p^{\text{alg}}, \sigma_q)$ where $\sigma_q \colon x \mapsto x^q$. Let $\mathcal{U}$ be a nonprincipal ultrafilter on the set $Q$ of all prime powers.

THEOREM (Hrushovski [7], Macintyre [8]). The difference field $\prod_{q \in Q} F_q / \mathcal{U}$ is a model of ACFA.

**8.5.** One can then show that the theory ACFA coincides with the set of sentences true in all but finitely many of the difference fields $F_q$. This is the analogue of Ax's theorem, which states that the theory of pseudofinite fields coincides with the set of sentences true in almost all finite fields. Hrushovski's proof gives more: it gives estimates on the size of finite definable sets. Before stating his result, let me remark that one can replace the scheme of axioms (3) of the theory ACFA by the apparently weaker scheme of axioms (3′).

(3′) If $U$ and $V$ are varieties *of dimension $d$* defined over $K$, with $V \subseteq U \times U^\sigma$ such that the projections of $V$ to $U$ and to $U^\sigma$ are generically onto, then there is a tuple $\bar{a}$ in $K$ such that $(\bar{a}, \sigma(\bar{a})) \in V$.

Let me explain why this (together with axiom schemes (1) and (2)) will give us an axiomatization of ACFA. The proof that every difference field $K$ embeds in a model of ACFA actually shows that every difference field embeds in a model $L$ of ACFA, all of whose elements are *transformally algebraic over $K$*. This means that given a tuple $\bar{a} \in L$, there is an integer $m$ such that $\text{acl}_\sigma(K \bar{a}) = K(\bar{a}, \dots, \sigma^m(\bar{a}))^{\text{alg}}$. Hence, every particular instance of axiom (3) is implied by finitely many instances of axiom (3′).

**8.6. Theorem** (Hrushovski). Let $F(\bar{X}, \bar{Z})$ and $G(\bar{X}, \bar{Y}, \bar{Z})$ be tuples of polynomials over $\mathbb{Z}$. There is a positive constant $C$ with the following property:

For all prime $p$ and power $q$ of $p$, and tuple $\bar{c}$ in $\mathbb{F}_p^{\text{alg}}$, if the equations $F(\bar{X}, \bar{c}) = 0$ and $G(\bar{X}, \bar{Y}, \bar{c}) = 0$ define varieties $U$ and $V$ satisfying the requirements of axiom (3′), then

$$\left| \text{Card}(\{\bar{a} \in \mathbb{F}_p^{\text{alg}} \mid (\bar{a}, \bar{a}^q) \in V\}) - cq^d \right| \le Cq^{d-1/2},$$

where

$$d = \dim(U) = \dim(V) \quad \text{and} \quad c = [K(V) : K(U)]/[K(V) : K(U^\sigma)]_{\text{ins}}.$$

This theorem is a strengthening of a theorem of Lang-Weil for the number of points of varieties defined over finite fields: take $U$ defined over $\mathbb{F}_q$, and $V$ the intersection of the diagonal with $U \times U$. The above formula then gives you an estimate on the number of points of $U$ with their coordinates in $\mathbb{F}_q$.

**8.7. Some applications of these theorems.** You may recall the proof of Ax that if $V$ is a variety defined over $\mathbb{C}$, and $f: V \to V$ is a morphism which is injective on the set $V(\mathbb{C})$ of points of $V$ with their coordinates in $\mathbb{C}$, then $f$ defines a bijection of $V(\mathbb{C})$ (i.e., $f$ is also surjective). This is done as follows: this statement will hold in $\mathbb{C}$ if and only if it holds in almost all algebraically closed fields of characteristic $p > 0$, if and only if it holds in almost all $\mathbb{F}_p^{\text{alg}}$, $p$ a prime. So let $V$ be a variety and $f: V \to V$ a morphism, both defined over $\mathbb{F}_p^{\text{alg}}$. Then for some $q = p^m$, they are defined over $\mathbb{F}_q$. Thus, for every $n \geq 1$, $f$ is a map $V(\mathbb{F}_{q^n}) \to V(\mathbb{F}_{q^n})$, is injective, and is therefore also surjective since $V(\mathbb{F}_{q^n})$ is finite. Because $\mathbb{F}_p^{\text{alg}} = \bigcup_n \mathbb{F}_{q^n}$, we get that $f$ defines a permutation of $V(\mathbb{F}_p^{\text{alg}})$.

COROLLARY. Let $B$ be a group of finite SU-rank defined in a model $K$ of ACFA, and assume that $f$ is a definable endomorphism of $B$. Then

$$[B : f(B)] = \text{Card}(\text{Ker}(f)).$$

PROOF. Without loss of generality, we may assume that $K$ is countable. Let $\varphi(\bar{x}, \bar{u})$, $\psi(\bar{x}, \bar{y}, \bar{u})$ be $\mathcal{L}$-formulas and $\bar{b} \in K$ be such that $\varphi(\bar{x}, \bar{b})$ defines $B$ and $\psi(\bar{x}, \bar{y}, \bar{b})$ defines the graph of $f$. By Theorem 8.4, the difference field $K$ embeds into an ultraproduct $\prod_q F_q/\mathcal{U}$ of the difference fields $F_q$. Take a representative $(\bar{b}_q)_q$ of $\bar{b}$. Then for almost all $q$ (in the sense of the ultrafilter $\mathcal{U}$) we have that $\varphi(\bar{x}, \bar{b}_q)$ defines a group $B_q$ in $F_q$, and $\psi(\bar{x}, \bar{y}, \bar{b}_q)$ defines an endomorphism $f_q$ of $B_q$. Note also that because $B$ is of finite SU-rank, there is an integer $m$ such that every element $\bar{a}$ of $B$ satisfies $\sigma^m(\bar{a}) \in \text{cl}_\sigma(\bar{b})(\bar{a}, \ldots, \sigma^{m-1}(\bar{a}))^{\text{alg}}$. Hence the same is also true for almost all $B_q$. By Theorem 8.6, the $B_q$ are therefore finite, and we obtain trivially that

$$[B_q : f_q(B_q)] = \text{Card}(\text{Ker}(f_q)).$$

Thus the same holds in $K$. Note that $\text{Ker}(f)$ is infinite if and only if the size of the $\text{Ker}(f_q)$ is unbounded.                                    ⊣

**8.8. Application to families of finite simple groups.** With the exception of the sporadic finite simple groups and the alternating groups $A_n$, the finite simple groups are members of infinite families. Some of these families are of the form $G(\mathbb{F}_q)$ for some linear algebraic group $G$ (for instance $SL_n(\mathbb{F}_q)$). Hence looking at $G(F)$ for $F$ a pseudofinite field, one gets uniformity results on the family $G(\mathbb{F}_q)$, $q$ a prime power.

Some of these families, however, do not originate directly from a simple algebraic group; their definition involves some automorphism of the field $\mathbb{F}_q$. They are the so-called twisted finite simple groups. And they become definable in the structure $(\mathbb{F}_p^{\text{alg}}, \sigma_q)$. Using Theorem 8.4, one then also gets uniformity results for these families.

Typical examples of uniformity results: we know that in a finite simple group $H$, every nontrivial conjugacy class $X$ generates the whole group. The uniformity gives a bound on the number $N$ such that $H = (X \cdot X^{-1})^N$.

**8.9. The Jacobi conjecture for difference fields.** Let $n \geq 1$, let $u_1(X_1, \ldots, X_n)$, $\ldots, u_n(X_1, \ldots, X_n) \in K[X_1, \ldots, X_n]_\sigma$ ($K$ a difference field, and we will assume it is a model of ACFA). The equations

$$u_1(x_1, \ldots, x_n) = \cdots = u_n(x_1, \ldots, x_n) = 0$$

define a $\sigma$-closed subset of $K^n$, and some of the irreducible components of this set will be of finite SU-rank. If $Y$ is an irreducible component of finite SU-rank of this set, let us define the *order of* $Y$ as $\mathrm{Sup}\{\mathrm{tr.deg}(\mathrm{cl}_\sigma(E\bar{a})/E) \mid \bar{a} \in Y\}$, where $E = \mathrm{acl}_\sigma(E)$ is such that the elements $u_1, \ldots, u_n$ are defined over $E$. The Jacobi conjecture gives an explicit bound $H$ on the order of the irreducible components of finite SU-rank. This bound is defined as follows: for each $k$ and $i$, define $h_k^i$ to be the order of $u_k$ when viewed as a difference polynomial in $X_i$ (so it equals $m$ if $X_i^{\sigma^m}$ appears in $u_k$ and $X_i^{\sigma^{m+1}}$ does not). One then sets

$$H = \max_{\theta \in \mathrm{Sym}(n)} \sum_{k=1}^{n} h_k^{\theta(k)}.$$

Hrushovski uses the fact that if an irreducible $\sigma$-closet set $Y$ has order $d$, then in a structure $F_q$, it will have approximately $cq^d$ points for some fixed constant $c$. Thus the order of $Y$ will be $\lim_q \log_q(\mathrm{Card}(Y(F_q)))$. One uses this remark to reduce the Jacobi conjecture to a problem about number of points in algebraic sets.

**8.10. Modular subgroups.** One can show that if one has an exact sequence

$$0 \to A \to B \to C \to 0$$

of definable groups (in a model $K$ of ACFA), then $B$ is modular (and stable) if and only if $A$ and $C$ are modular (and stable). Hence the study of modular definable subgroups of an algebraic group reduces to the study of definable subgroups of simple algebraic groups, i.e., of $\mathbb{G}_a(K)$, $\mathbb{G}_m(K)$, or of a simple abelian variety.

In characteristic 0, one can show that *no* proper definable subgroup of $\mathbb{G}_a(K)$ is modular. Indeed, it is fairly easy to show that such a group is commensurable with one defined by an equation of the form $\sum_{i=0}^{n} a_i \sigma^i(x) = 0$, with $a_i \in K$ (recall that two definable groups $G_1$ and $G_2$ are *commensurable* if $[G_1 : G_1 \cap G_2]$ and $[G_2 : G_1 \cap G_2]$ are both finite). Furthermore, the operator $\sum_{i=0}^{n} a_i \sigma^i$ can be written as a composition of operators of the form $\sigma - b$, and the sets defined by the equation $\sigma(x) - bx$ are nonorthogonal to the fixed field.

Let $A$ be an abelian variety defined over the fixed field $F$ of the model $K$ of ACFA. Then $\sigma(A) = A$, and therefore, if $f(T) = \sum_{i=0}^{n} a_i T^i \in \mathbb{Z}[T]$, then the equation

$$f(\sigma)(x) = \sum_{i=0}^{n} [a_i]\sigma^i(x) = 0$$

defines a subgroup of $A(K)$. (Here, $[a_i]x$ denotes $x + \cdots + x$ $a_i$ times). We will denote this group by $\mathrm{Ker}(f(\sigma))$.

A beautiful result of Hrushovski states that in characteristic 0, $f(T)$ is relatively prime to all cyclotomic polynomials $T^m - 1$ if and only if $\mathrm{Ker}(f(\sigma))$ is modular and therefore stable. (He actually has a complete characterization of the modular definable subgroups of an arbitrary abelian variety.)

A similar statement holds for definable subgroups of $\mathbb{G}_m(K)$: if $f(T) = \sum_i a_i T^i \in \mathbb{Z}[T]$, then the equation $\prod \sigma^i(x^{a_i}) = 1$ defines a subgroup of $\mathbb{G}_m(K)$ (denoted $\mathrm{Ker}(f(\sigma)))$, and this subgroup is modular if and only if $f(T)$ is relatively prime to all cyclotomic polynomials $T^m - 1, m \geq 1$.

**8.11. Application to the Manin-Mumford conjecture.** The Manin-Mumford conjecture is a conjecture in number theory of which various versions have been proved by various people. Here I will only state the version of which Hrushovski gives a proof using difference fields. It is weaker than the full Manin-Mumford conjecture because of the restriction on the field of definition, but is stronger in the sense that the constant $M$ below can be effectively computed from the data. If $G$ is a group, let us denote by $\mathrm{Tor}(G)$ the set of torsion elements of $G$. I will certainly not give the full proof, but will try to indicate the main steps of the proof and some of its ingredients.

THEOREM. Let $A$ be a commutative algebraic group defined over some number field $k$ ($=$ finite extension of $\mathbb{Q}$), and let $X$ be a subvariety of $A$. Then

$$X \cap \mathrm{Tor}(A) = \bigcup_{i=1}^{M} a_i + \mathrm{Tor}(A_i)$$

where the $A_i$ are group subvarieties of $A$. The number $M$ is bounded above by $c \deg(X)^e$ where $c$ and $e$ are constants depending on $A$, and $\deg(X)$ is the degree of $X$ given some embedding of $A$ (and of $X$) in projective space.

**8.12. A first idea for a proof.** Show that there is $\sigma \in \mathcal{G}\mathrm{al}(k^{\mathrm{alg}}/k)$ and a functional equation $f(\sigma)(\bar{x}) = 0$ valid on $\mathrm{Tor}(A)$ and such that, whenever $K$ is a model of ACFA extending $(k^{\mathrm{alg}}, \sigma)$, then the equation $f(\sigma)(\bar{x}) = 0$ defines a modular subgroup of $A(K)$.

Assume that we have done that.

(1) We get the qualitative version of the result because of modularity. Indeed, let $B$ be the subgroup of $A(K)$ defined by the equation $f(\sigma)(\bar{x}) = 0$. Then by the results we got above, $X \cap B$ is a Boolean combination of cosets of quantifier-free definable subgroups of $B$. We can write $X \cap B$ as a finite union of sets of the form $(a + C) \setminus U$ where $C$ is a $\sigma$-closed subgroup

of $B$, irreducible as a $\sigma$-closed set, and $U$ is a union of cosets which are strictly contained in $a + C$. The $\sigma$-closure of $(a + C) \setminus U$ equals $a + C$ and must also be contained in the $\sigma$-closed set $X \cap B$. This shows that $X \cap B$ is in fact a finite union of cosets of definable subgroups $B_i$ of $B$. Write

$$X \cap B = \bigcup_{i=0}^{n} a_i + B_i.$$

Now, if $a_i + B_i$ intersects $\mathrm{Tor}(A)$, then we may take $a_i \in \mathrm{Tor}(A)$. Note also that $a_i + B_i \subset X$ implies that $a_i + A_i \subset X$ where $A_i$ is the Zariski closure of the subgroup $B_i$ of $A$ and therefore is an algebraic subgroup of $A$. Hence we have that $a_i + \mathrm{Tor}(A_i) \subset X$, and this gives the result.

(2) If we have explicit bounds on the degree of $f$ and the absolute values of its coefficients, then we get an explicit bound on the number of irreducible components of the $\sigma$-closed set $X \cap B$ and hence on the number $M$ of cosets.

Unfortunately, this strategy needs to be slightly modified. First of all, the commutative algebraic group $A$ may have a *vector subgroup*, i.e., an algebraic subgroup $V$ which is isomorphic to $\mathbb{G}_a^n$ for some $n$. As we saw above, no definable subgroup of $\mathbb{G}_a^n(K)$ is modular. So the first step of the proof is to show that it is enough to prove the result for $A/V$ where $V$ is the maximal vector subgroup of $A$. This is done using an easy algebraic result and a more complicated model-theoretic one showing that if $B$ is a definable subgroup of $A(K)$ which is such that $B/B \cap V$ is modular, then the intersection of $B$ with any set is of a special form (see below in Subsection 8.15).

So one reduces the proof of the theorem to the case where $A$ has no algebraic subgroups isomorphic to $\mathbb{G}_a$ (such an algebraic group is called a *semiabelian variety*).

Also, it turns out that finding explicit bounds on coefficients and degree of the equation is not effective. One bypasses this difficulty by looking first at $\mathrm{Tor}_{p'}(A)$, the prime-to-$p$ torsion subgroup of $A$, proving the desired result there, and then using another prime $\ell$. This produces explicit but very ugly bounds on the number $n$ of cosets.

More precisely, we get a bound for the number of cosets in $X \cap \mathrm{Tor}_{p'}(A)$, and this bound only depends on the degree of $X$. Hence, it also works for $(a + X) \cap \mathrm{Tor}_{p'}(A)$ for any $a \in \mathrm{Tor}(A)$. Now one uses that $\mathrm{Tor}_{p'}(A) + \mathrm{Tor}_{\ell'}(A) = \mathrm{Tor}(A)$. But there is definitely some work involved. For details, please see [6].

**8.13. Why do we need a bound on the coefficients of the equation?** Let $f(T) = \sum_{i=0}^{\ell} m_i T^i \in \mathbb{Z}[T]$. We are interested in the number of irreducible components of the $\sigma$-closed set $X \cap \mathrm{Ker}(f(\sigma))$, and we can find a bound on this number as follows. We first look at the Zariski closed sets $Y = (X \times X^\sigma \times \cdots \times X^{\sigma^\ell})$ and $C = \{(x_0, \ldots, x_\ell) \in A^{\ell+1} \mid \sum_{i=0}^{\ell} [m_i] x_i = 0\}$. Then the number of irreducible components of $Y \cap C$ is bounded by $\deg(X)^{\ell+1} \deg(C)$

where the degree is computed via a certain embedding of $A$ in projective space. Unfortunately, the degree of $C$ will depend on the values of the $|m_i|$ (and increase if they do, of course) which means that already to know the number of irreducible components of $Y \cap C$, we need to know things about $f(T)$.

**8.14. Existence of the equation and some bounds.** Here Hrushovski chooses a prime $p$ of good reduction (for $A$). Grosso modo, this means that, reducing modulo $p$ the equations defining $A$, one gets an algebraic group $\bar{A}$ defined over some finite field $\mathbb{F}_q$ and which resembles $A$. Moreover, let $L_p$ be the field generated over $k$ by the elements of $\mathrm{Tor}_{p'}(A)$. Then $L_p$ is an unramified extension of the field $k$, i.e., reduction "mod $p$": $\mathcal{O}_k \to \mathbb{F}_q$ extends to a homomorphism $\mathcal{O}_{L_p} \to \mathbb{F}_p^{\mathrm{alg}}$, and this homomorphism defines a 1-1 map on $\mathrm{Tor}_{p'}(A)$ (with values in $\mathrm{Tor}_{p'}(\bar{A})$).

The automorphism $x \mapsto x^q$ of $\mathbb{F}_p^{\mathrm{alg}}$ lifts to an automorphism of $L_p$ fixing $k$, and one takes for $\sigma$ any automorphism of $\mathbb{Q}^{\mathrm{alg}}$ extending this automorphism.

It then suffices to find the functional equation on $\bar{A}$ (because reduction mod $p$ is injective on the torsion). Observing that if

$$0 \longrightarrow A_1 \longrightarrow A_3 \longrightarrow A_2 \longrightarrow 0$$

is a short exact sequence of (connected) commutative algebraic groups, then showing $\mathrm{Tor}(A_3)$ contains $\mathrm{Tor}(A_1)$ and projects into $\mathrm{Tor}(A_2)$, reduces to finding such an equation (and bounds on its coefficients and degree) for each of the simple factors of $\bar{A}$. For the abelian ones, its existence and bound on the degree and absolute value of the coefficients are given by a result of Weil. For products of copies of $\mathbb{G}_m$, it is more or less $\sigma(x) - x^q$, and $\mathbb{G}_a(K)$ has no torsion elements, so we do not need to look at what happens in $\mathbb{G}_a(\mathbb{F}_p^{\mathrm{alg}})$.

We have therefore taken care of the $p'$-torsion, and furthermore have shown that there is an effective bound on the number $n$ such that

$$\mathrm{Tor}_{p'}(A) \cap X = \bigcup_{i=1}^{n} a_i + \mathrm{Tor}_{p'}(A_i)$$

for some algebraic subgroups $A_i$ of $A$ and elements $a_1, \ldots, a_n$. Take now another prime $\ell$ of good reduction. Proceeding as above, one gets an automorphism $\tau$ of $\mathbb{Q}^{\mathrm{alg}}$ such that $\tau$ satisfies a functional equation $g(\tau)$ on $\mathrm{Tor}_{\ell'}(A)$ and therefore also on $\mathrm{Tor}_p(A)$ (the $p$-component of $\mathrm{Tor}(A)$). Let $M_p = k(\mathrm{Tor}_p(A))$. Then a result of Serre tells us that $L_p \cap M_p$ is a *finite* extension of $k$, say of degree $m$. Hence there is an automorphism $\rho$ of $L_p M_p$ which extends $\sigma^m$ on $L_p$ and $\tau^m$ on $M_p$. Note the following: write $f(T) = a \prod(T - \alpha_i)$ where $a \in \mathbb{Z}$, the $\alpha_i$ are in $\mathbb{C}$. Then $\mathrm{Ker}(f(\sigma))$ is modular if and only if none of the $\alpha_i$ is a root of unity. Also, let $h(T) = \prod(T - \alpha_i^m)$, and choose $b \in \mathbb{Z}$ such that $bh(T) \in \mathbb{Z}[T]$; then $bh(\sigma^m)$ vanishes on $\mathrm{Tor}_{p'}(A)$ and is relatively prime to all cyclotomic polynomials, hence defines a modular subgroup of $A(K)$. Reasoning similarly with $\tau$, we get that there is a modular subgroup $B$ of some model of ACFA extending $(L_p M_p, \rho)$ which contains Tor($A$). This gives us the qualitative version of Manin-Mumford conjecture: we have shown that Tor($A$) $\cap X$ is a finite union of cosets of the torsion subgroups of some connected algebraic subgroups of $A$. This fact is used in the proof.

Unfortunately, we do not know a priori that the constant $m$ of Serre's result has an effective bound. This means that we do not know a bound on the complexity of the set $B$ defining our modular subgroup and hence cannot derive a bound on the number of irreducible components of the $\sigma$-closed set $B \cap X$. However, we do have explicit bounds (depending on $p$ and $\ell$) on the number of cosets appearing in the decompositions of $\text{Tor}_{p'}(A) \cap X$ and $\text{Tor}_{\ell'}(A) \cap X$, and this will be enough to apply the strategy alluded to at the end of Subsection 7.12.

**8.15. The reduction to the semiabelian variety case.** We will state here the two results which are needed to reduce the problem from the commutative algebraic group $A$ to the semiabelian variety $A/V$. We will not state the model-theoretic result in its full generality, only the particular case that is of interest to us.

DEFINITION. Let $A$ be a commutative algebraic group, $V$ its maximal vector group. We call a subvariety $X$ of $A$ *special* if $X = Y + C$ where $Y$ is a subvariety of $V$ and $C$ is a coset of a (connected) algebraic subgroup $E$ of $A$.

We call a definable subset $D$ of $A(K)$ *special* if $X = Y + C$ where $Y$ is a definable subset of $V(K)$, $C$ is a coset of a definable subgroup of $A(K)$.

**8.16. Proposition.** Let $A$ be a commutative algebraic group and $V$ its maximal vector subgroup, $\pi: A \to A/V$. Let $B$ be a definable subgroup of $A(K)$ of finite SU-rank and assume that $\pi(B)$ is modular. Then every definable subset of $B$ is a Boolean combination of special sets.

The proof uses the following crucial ingredients:

(1) The fact that any definable subset of $\text{Fix}(\sigma)^n$ is definable with parameters from $\text{Fix}(\sigma)$ (see Proposition 5.3).
(2) The stability and the modularity of $\pi(B)$ with the structure induced by $K$.
(3) The fact that $B \cap V$ is strongly related to $\text{Fix}(\sigma)$. (In fact, there is a definable bijection between $B \cap V$ and $\text{Fix}(\sigma)^k$.)
(4) The fact that any definable subset of $(V \cap B) \times \pi(B)$ is a Boolean combination of "rectangles", i.e., sets of the form $U_1 \times U_2$ where $U_1, U_2$ are definable and included in $B \cap V$ and $\pi(B)$, respectively.
(5) If $Y$ is a definable subset of $B$, then there is a group $H$, contained in $(Y \cdot Y^{-1})^n$ for some $n$ (and hence definable), such that $YH/H$ is finite. This is a property of groups of finite SU-rank.

**8.17. Corollary.** Let $B$ be as above and $X$ a subvariety of $A$. Then $X \cap B$ is contained in a finite union of finitely many special varieties which are contained in $X$.

PROOF. Go to the Zariski closure of $X \cap B$.                              ⊣

**8.18. Lemma.** Let $A$ be a commutative algebraic group, $T$ the group of torsion points of $A$ (or the group of torsion points of order prime to $p$, for some prime $p$), and assume that $X \cap T$ is contained in the union of the special subvarieties $D_i$ of $X$, $i = 1, \ldots, M$. Then the Zariski closure of $X \cap T$ is the union of at most $M$ cosets of connected algebraic subgroups of $A$.

PROOF. It is enough to show that if $D$ is a special subvariety of $X$, then the Zariski closure of $D \cap T$ is a coset of a group subvariety of $A$.

Write $D = C + Y$ where $C$ is a coset of the connected group variety $E$ and $Y$ is a subvariety of $V$. We will first show that we may assume that $Y$ is contained in an algebraic subgroup $V_1$ of $V$ which intersects $E$ in $(0)$. Indeed, because $V$ is a vector space, there is a definable endomorphism $\pi \colon V \to V$ with $\pi^2 = \pi$ and $\mathrm{Ker}(\pi) = V \cap E$. Then $E + Y = E + \pi(Y)$ because $\mathrm{Ker}(\pi) \subseteq E$, and $C = C + E$ because $C$ is a coset of $E$. Therefore,

$$D = C + Y = C + E + Y = C + E + \pi(Y) = C + \pi(Y),$$

and we may replace $Y$ by $\pi(Y) \subseteq \pi(V) = V_1$, and $V_1 \cap E = (0)$.

Fix $d_0 \in D \cap T$ and for $d \in D \cap T$, define $f(d) = d - d_0$. Then $f(d) \in (E + V_1) \cap T$. Write $f(d) = e + y$ with $e \in E$, $y \in V_1$. Since $f(d)$ is a torsion element and $V_1 \cap E = (0)$, we get that necessarily $y = 0$ ($V$ has no torsion elements) so that $f(d) \in E$, and $d \in E + d_0$. Hence,

$$D \cap T = (E + d_0) \cap T = (E \cap T) + d_0.$$

The only thing remaining to be shown, is that the Zariski closure $B$ of $E \cap T$ is a connected subgroup of $A$. Let $B^0$ be the connected component of $B$. Because $E \cap T$ is divisible, $E \cap T$ has no subgroup of finite index which implies that $B^0 \cap T = B$ so that $B = B^0$.                              ⊣

**8.19. Exercise 16.** Let $(E, \sigma)$ be an algebraically closed difference field, let $m \geq 1$, and let $(L, \tau)$ be a difference field extending $(E, \sigma^m)$. For each $i = 1, \ldots, m - 1$, choose $L_i$ realizing $\sigma^i(\mathrm{tp}_{\mathrm{ACF}}(L/E))$ linearly disjoint from the composite field of $L_0 = L, \ldots, L_{i-1})$ over $E$. Let $f_0 = id_L$ and for $i = 1, \ldots, m - 1$, let $f_i \colon L \to L_i$ be an isomorphism extending $\sigma^i$. For $i = 0, \ldots, m - 2$, define $\sigma_i \colon L_i \to L_{i+1}$ by

$$\sigma_i = f_{i+1} f_i^{-1}$$

and define $\sigma_{m-1} \colon L_{m-1} \to L_0$ by

$$\sigma_{m-1} = \tau f_{m-1}^{-1}.$$

(1) Show that the $\sigma_i$ agree with $\sigma$ on $E$.
(2) Show that $\sigma_{m-1} \ldots \sigma_0 = \tau$.

(3) Show that there is a difference field $(M, \sigma)$ containing $(E, \sigma)$ and such that $M$ contains $L$, and $\sigma^m$ agrees with $\tau$ on $L$.

(4) Deduce that if $K$ is a model of ACFA, then $(K, \sigma^m \phi^n)$ is also a model of ACFA $(m \geq 1, n \in \mathbb{Z})$.

(5) Deduce that if $K$ is a model of ACFA and $F = \text{Fix}(\sigma^m \phi^n)$, then $F$ is a pseudofinite field.

(6) Let $F$ be as in (5). Show that every definable (in $K$) subset of $F^k$ is definable with parameters from $F$. If, furthermore, $m = 1$, then show that the structure induced on $F$ is the pure field structure. (Repeat the proof of Proposition 5.3. This result does not extend to the case $m > 1$, as $\sigma$ defines then an automorphism of $F$ which is not definable in the pure field language.)

REFERENCES

Algebraic results:

[1] RICHARD M. COHN, *Difference algebra*, Interscience Publishers John Wiley & Sons, New York-London-Sydney, 1965.

[2] SERGE LANG, *Introduction to algebraic geometry*, Addison-Wesley Publishing Co., Inc., Reading, Mass., 1973, Fourth printing.

Model theoretic results on difference fields:

[3] ANGUS MACINTYRE, *Generic automorphisms of fields*, **Annals of Pure and Applied Logic**, vol. 88 (1997), no. 2–3, pp. 165–180, Joint AILA-KGS Model Theory Meeting (Florence, 1995).

[4] ZOÉ CHATZIDAKIS and EHUD HRUSHOVSKI, *Model theory of difference fields*, **Transactions of the American Mathematical Society**, vol. 351 (1999), no. 8, pp. 2997–3071.

[5] Z. CHATZIDAKIS, E. HRUSHOVSKI, and Y. PETERZIL, *Model theory of difference fields, II: Periodic ideals and the trichotomy in all characteristics*, preprint 1999, **Proceedings of The London Mathematical Society**, vol. 85 (2002), pp. 257–311.

[6] EHUD HRUSHOVSKI, *The Manin-Mumford conjecture and the model theory of difference fields*, **Annals of Pure and Applied Logic**, vol. 112 (2001), no. 1, pp. 43–115.

[7] EHUD HRUSHOVSKI, *The first-order theory of the Frobenius*, preprint, 1996.

[8] A. MACINTYRE, *Nonstandard Frobenius*, in preparation.

The papers [5] to [7] definitely fall into the category of "further advanced reading". Some of the easy parts of [6] appear in survey papers, see below.

Related model-theoretic results on finite fields:

[9] JAMES AX, *The elementary theory of finite fields*, **Annals of Mathematics (2)**, vol. 88 (1968), pp. 239–271.

[10] E. HRUSHOVSKI, *Pseudo-finite fields and related structures*, manuscript, 1991.

[11] EHUD HRUSHOVSKI and ANAND PILLAY, *Groups definable in local fields and pseudo-finite fields*, **Israel Journal of Mathematics**, vol. 85 (1994), no. 1–3, pp. 203–262.

[12] E. HRUSHOVSKI and A. PILLAY, *Definable subgroups of algebraic groups over finite fields*, **Journal für die Reine und Angewandte Mathematik**, vol. 462 (1995), pp. 69–91.

The proofs of many of the results of the first few sections of these notes are essentially translations of proofs appearing in [10]. See also [11]. In [12], one uses results of [11] to obtain very pretty results on groups definable in finite fields.

Other model-theoretic results:

[13] L. VAN DEN DRIES and K. SCHMIDT, *Bounds in the theory of polynomial rings over fields. An approach*, **Inventiones Mathematicae**, vol. 76 (1984), no. 1, pp. 77–91.

[14] E. HRUSHOVSKI and A. PILLAY, *Weakly normal groups*, **Logic colloquium '85 (Orsay, 1985)**, North-Holland, Amsterdam, 1987, pp. 233–244.

[15] BYUNGHAN KIM and ANAND PILLAY, *Simple theories*, **Annals of Pure and Applied Logic**, vol. 88 (1997), no. 2–3, pp. 149–164, Joint AILA-KGS Model Theory Meeting (Florence, 1995).

[16] ANAND PILLAY, *Definability and definable groups in simple theories*, **The Journal of Symbolic Logic**, vol. 63 (1998), no. 3, pp. 788–796.

Stability is a vast subject with an abundant literature. Here are some books for the interested reader, but there are many others.

[17] STEVEN BUECHLER, *Essential stability theory*, Perspectives in Mathematical Logic, Springer-Verlag, Berlin, 1996.

[18] ANAND PILLAY, *An introduction to stability theory*, The Clarendon Press Oxford University Press, New York, 1983.

[19] FRANK O. WAGNER, *Simple theories*, Kluwer Academic Publishers, Dordrecht, 2000.

Survey papers:

[20] E. BOUSCAREN, *Théorie des modèles et conjecture de Manin-Mumford*, [d'après Ehud Hrushovski], Séminaire Bourbaki vol. 1999/2000, *Astérisque*, no. 276 (2002), pp. 137–159.

[21] ZOÉ CHATZIDAKIS, *Groups definable in ACFA*, **Algebraic model theory (Toronto, ON, 1996)** (B. Hart et al., editors), Kluwer Academic Publishers, Dordrecht, 1997, pp. 25–52.

[22] ZOÉ CHATZIDAKIS, *A survey on the model theory of difference fields*, **Model theory, algebra, and geometry** (Steinhorn, Haskell, Pillay, editors), MSRI Publications, vol. 39, Cambridge University Press, 2000, pp. 65–96.

[23] ANAND PILLAY, *ACFA and the Manin-Mumford conjecture*, **Algebraic model theory (Toronto, ON, 1996)**, (B. Hart et al., editors) Kluwer Academic Publishers, Dordrecht, 1997, pp. 195–205.

After these notes were written, progress was made by several people. Some results were simplified or generalised (Proposition 6.17 in [24], the dichotomy Theorem 8.2 in characteristic 0 in [25]). Other results of interest appear in [26], [27] and [28].

[24] FRANK O. WAGNER, *Some remarks on one-basedness*, **The Journal of Symbolic Logic**, vol. 69 (2004), no. 1, pp. 34–38.

[25] ANAND PILLAY AND MARTIN ZIEGLER, *Jet spaces of varieties over differential and difference fields*, **Selecta Mathematica. New Series**, vol. 9 (2003), no. 4, pp. 579–599.

[26] THOMAS SCANLON, *Diophantine geometry of the torsion of a Drinfeld module*, **Journal of Number Theory**, vol. 97 (2002), no. 1, pp. 10–25.

[27] RICHARD PINK AND DAMIAN ROESSLER, *On Hrushovski's proof of the Manin-Mumford conjecture*, **Proceedings of the International Congress of Mathematicians, (Beijing, 2002)**, vol. I, pp. 539–546, Higher Education Press, Beijing, 2002.

[28] ANAND PILLAY, *Mordell-Lang conjectures for function fields in characteristic zero, revisited*, **Compositio Mathematica**, vol. 140 (2004), no. 1, pp. 64–68.

UFR DE MATHÉMATIQUES
UNIVERSITÉ PARIS 7 - CASE 7012
2, PLACE JUSSIEU
75251 PARIS CEDEX 05, FRANCE
*E-mail*: zoe@logique.jussieu.fr

# Index for *Model theory of difference fields*

# SOME COMPUTABILITY-THEORETIC ASPECTS
# OF REALS AND RANDOMNESS

RODNEY G. DOWNEY

**Abstract.** We study computably enumerable reals (i.e., their left cut is computably enumerable) in terms of their spectra of representations and presentations. Then we study such objects in terms of algorithmic randomness, culminating in some recent work of the author with Hirschfeldt, Laforte, and Nies concerning methods of calibrating randomness.

§1. **Introduction.** We study reals $\alpha$, $0 < \alpha < 1$, unless otherwise specified and for convenience no real will be rational. This convention allows us to give uniform proofs of many results which would otherwise split into cases of whether the real at hand was rational or not. In particular if we have a sequence of rationals converging to a real $\alpha$ then this sequence will be infinite, and furthermore every such real will have a unique dyadic expansion.

Much of modern computability theory is concerned with understanding the computational complexity of sets of *positive integers*, yet even in the original paper of Turing [64], a central topic is interest in effectiveness considerations for *reals*. Of particular interest to computable analysis (e.g., Weihrauch [66], Pour-El [52], Pour-El and Richards [53], Ko [39]) and to algorithmic information theory (e.g., Chaitin [10], Calude [6], Martin-Löf [51], Li-Vitanyi [48]) is the collection of *computably enumerable reals*. These are the reals $\alpha$ such that the lower cut $L(\alpha)$ consisting of rationals less than $\alpha$ forms a computably enumerable set.

The first part of these notes consists of an analysis of the basic ways that we present reals and a clarification of the relationship between such presentations and degree classes. In particular we will look at the recent work of Calude, Coles, Hertling, Khoussainov [7], Downey [12], Downey and Laforte [27], Ho [34], and Wu [68], as well as older work of Soare [58] and others.

Research partially supported by the Marsden Fund of New Zealand. These notes are based upon a short course of lectures given in the fall of 2000 at the University of Notre Dame. The author thanks all the logicians there for their hospitality and support. He also thanks Andrew Arana for his skillful notetaking.

Our main goal is to look at algorithmic randomness, especially with respect to computably enumerable reals. To this end we will next introduce the basic approaches to the study of algorithmic randomness, both topological notions, as in Martin-Löf randomness, and compressibility notions such as Chaitin-Kolmogorov randomness. We begin by looking at these notions for finite strings and then proceed to reals.

Finally, we will look at some recent material of the calibration of relative randomness using notions such as Solovay reducibility and some new reducibilities, $rK$ and $sw$.

Notation is more or less standard and follows Soare [61]. As these notes are aimed at graduate students, and one learns from actively engaging in the material, we will not always provide complete proofs, but will always provide sketches, referring the reader to the appropriate paper when necessary.

Of course, we identify reals with their characteristic functions when considered as dyadic expansions. (Remember, no real is rational.) Hence, if I write $\alpha = .A$, I mean that $\alpha = \sum_{n \in A} 2^{-n}$.

§2. **Reals, computable or otherwise.** What is a real? Our first view of this question is that a real is a cut. Let $\alpha$ be a real. Then by $L(\alpha)$ we mean the left cut of $\alpha$;

$$L(\alpha) = \{q \in \mathbb{Q} : q < \alpha\}.$$

We may approximate $\alpha$ via a Cauchy sequence, viz., $\alpha = \lim_s q_s$.

What is a computable real? Here are three guesses based on the Cauchy definition.

   (i) $\alpha$ is the limit of a computable sequence of rationals.
   (ii) $\alpha$ is the limit of a computable monotone increasing sequence of rationals.
   (iii) $\alpha = .A$ for some computable set $A$. (Here we consider $A$ as identified with its characteristic function so that this is the dyadic expansion of $\alpha$.)

Now it is a fact that $(iii) \rightarrow (ii) \rightarrow (i)$ but none of the implications can be reversed. If (iii) holds (and note that we could equally have used a decimal expansion), then there is an algorithm $M$ allowing us to compute $L(\alpha)$; namely, given $n$, we can compute $q_n = .A \quad n + 1$ so that $|\alpha - q_n| < 2^{-n}$, so that given $q$ we can calculate $q_n$s until either $q$ appears in $L(\alpha)$ or it becomes bounded away from $\alpha$. We cannot guarantee this in either (i) or (ii) since we lack an effective radius of convergence. Suppose that we call a real $\alpha = .A$ a computable real, if $A$ is a computable set. The following is implicit in Turing's original paper.

THEOREM 1 (Turing). *$\alpha$ is a computable real iff it is the limit of a computable sequence of rationals $q_i : i \in B$ and there is a computable algorithm $M$ so that for all $n$,*

$$|\alpha - q_{M(n)}| < 2^{-n}.$$

The proof is left as an exercise with the hint for the 'only if' direction being that since the real is not rational there is always another 0 in its dyadic expansion. Note that we have not actually proved that (ii) and (iii) are different yet, only that they seem different. We look at this now.

DEFINITION 1. *We call a real $\alpha$ computably enumerable (also sometimes, left computable, left c.e. semi-computable in the literature) iff $L(\alpha)$ is computably enumerable.*

We will need a technical notion whose use is crucial in later investigations, especially in terms of randomness. A set $A \subseteq 2^{<\omega}$ of strings is called *prefix-free* iff for all $\sigma \in A$, and all $\tau$ with $\sigma$ an initial segment of $\tau$, $\tau \notin A$. Prefix-free sets are considered for technical reasons since if a set $A$ is prefix free then, as we soon see, by Kraft's inequality, we know that $\sum_{n \in A} 2^{-|n|}$ converges and conversely.

THEOREM 2 (Calude, Khoussainov, Hertling, Wang [8], Soare [58]). *The following are equivalent.*

(i) *$\alpha$ is the limit of a computable enumerable monotone increasing (in the real ordering) sequence of rationals.*

(ii) *$\alpha$ is computably enumerable.*

(iii) *There is an infinite computably enumerable prefix-free set $A$ with $\alpha = \sum_{n \in A} 2^{-|n|}$.*

(iv) *There is a computable prefix-free set $A$ such that $\alpha = \sum_{n \in A} 2^{-|n|}$.*

(v) *There is a computable function $f(x, y)$ of two variables such that*

(va) *if, for some $k, s$ we have $f(k, s + 1) = 0$ yet $f(k, s) = 1$ then there is some $k' < k$ such that $f(k', s) = 0$ and $f(k', s + 1) = 1$;*

(vb) *$\alpha = \cdot a_1 a_2 \ldots$ is a dyadic expansion of a with $a_i = \lim_s f(i, s)$.*

(vi) *There is a computable increasing sequence of rationals with limit $\alpha$.*

The reader should be aware of the two orderings at work here. In (i) the rationals are coded and the sequence of codes computably enumerable. It is possible to have the sequence "increasing" as a sequence of rationals in the real ordering yet as codes they could be decreasing. For (v) we mean that there is a computable function $g : \omega \mapsto \mathbb{Q}$ with $\alpha = \lim_s g(s)$ and the range of $g$ a computable set of (codes of) rationals.

It is important that the reader realize that we are *not* defining a c.e. real to be $.A$ for some c.e. set $A$. Define a real $\alpha$ to be *strongly c.e.* if there is a c.e. set $A$ such that $\alpha = .A$. It is easy to use the characterization above (specifically (iv)) to construct a c.e. real that is not strongly c.e. (a theorem of Soare [58]). Specifically, we need to satisfy the requirement

$$R_j : \alpha \neq .W_e.$$

The idea is very simple. Devote positions $2e$ and $2e + 1$ to $R_e$. We initially set $A(2e + 1) = 1, A(2e) = 0$. If ever $2e + 1 \in W_e$, make $A(2e + 1) = 0$ and $A(2e) = 1$.

Notice that every strongly c.e. real is c.e. but that if $A$ is c.e. and not computable, then $\alpha = .A$ is c.e. and cannot be computable.

The sets $A$ which have enumerations satisfying (v) we call *nearly c.e.* and occupy a special place in our investigations.

None of the proofs is difficult. Why does (ii) imply (vi)? We need to replace $q_0, q_1, \ldots$ with a computable enumeration with the same limit. Let $<_R$ denote the real ordering. We simply find a sequence of rationals with $q_n <_R r_n <_R q_{n+1}$ and such that the code of $r_{n+1}$ exceeds that of $r_n$, which is possible by the density of the rationals. The sequence $r_n$ so obtained has the same limit as the $q_i$ and is increasing in Gödel number. All of the remaining implications are left to the reader, save the ones involving prefix free sets. For these results, we use a very important theorem called Kraft's inequality.

THEOREM 3 (Kraft).

(i) *If $A$ is prefix free then $\sum_{n \in A} 2^{-|n|} \leq 1$.*

(ii) *(sometimes called Kraft-Chaitin, or Chaitin simulation) Let $d_1, d_2, \ldots$ be a collection of lengths, possibly with repetitions. Then $\sum 2^{-d_i} \leq 1$ iff there is a prefix-free set $A$ with members $\sigma_i$ and $\sigma_i$ has length $d_i$. Furthermore from the sequence $d_i$ we can effectively compute the set $A$.*

PROOF. One direction of Kraft-Chaitin is clear. This is because of the topological correspondence $\Delta: [\sigma] \mapsto [0.\sigma, 0.\sigma + 2^{-|\sigma|})$ taking string $\sigma$ to an interval of size $2^{-|\sigma|}$, gives a correspondence betweeen a set of disjoint intervals in $[0, 1)$ and a prefix-free set.

We begin by discussing the original non-effective proof of Kraft and then look at the effectivization of Chaitin. We are given the lengths $\{d_i : i \in \mathbb{N}\}$ in some random order. But suppose that we re-arrange these lengths in increasing order, say $l_1 \leq l_2 \leq \ldots$. Then we can easily choose disjoint intervals $I_j$, with the right end-point of $I_n$ as the left endpoint of $I_{n+1}$ and the length of $I_{n+1}$ being $2^{l_{n+1}}$. Then we can again use the correspondence by setting $[\sigma_n] = \Delta^{-1}(I_n)$.

Now in the case that the intervals are not presented in increasing order, how to effective this process? The following idea was suggested by Joe Miller. At each stage $n$ the idea is that we will have a mapping $d_i \mapsto [\sigma_i], |\sigma_i| = d_i$, together with a binary string $x[n] = .x_1 x_2 \ldots x_m$ representing the length $1 - \sum_{j \leq n} 2^{-d_j}$, and ask that for each 1 in the expansion that there is a string of precisely that length in $2^{<\omega} - \{\sigma_j : j \leq n\}$.

To continue the induction, at stage $n + 1$, when a new length $d_{n+1}$ enters, if position $x_{d_{n+1}}$ is a 1, then we can find the corresponding string $\tau_{d_{n+1}}$ in $2^{<\omega} - \{\sigma_j : j \leq n\}$ and set $\sigma_{n+1} = \tau_{d_{n+1}}$. Then of course we make $x_{d_{n+1}} = 0$ in $x[n + 1]$.

If position $x_{d_{n+1}}$ is a 0, find the largest $j < d_{n+1}$ with $x_j = 1$, find the lexicographically least string $\tau$ extending $\tau_j$ of length $d_{n+1}$, let $\sigma_{n+1} = \tau$, and let $x[n+1] = x[n] - .v$ where $v$ is the string which is zero except for 1 in position $d_{n+1}$.

Notice that nothing changes in $x[n+1]$ from $x[n]$ except in positions $j$ to $d_{n+1}$, and these all change to 1, with the exception of $x_j$ which changes to 0. Since $\tau$ was chosen as the lexicographically least string in the cone $[\tau_j]$, there will be corresponding strings in $[\tau_j]$ of lengths $j - 1, \ldots, d_{n+1}$, as required to complete the induction.                                                                  ⊣

One way to think of the effective version of Kraft's inequality, the so-called Kraft-Chaitin theorem, is the following.

*We are effectively given a set of "requirements"* $\langle n_k, \sigma_k \rangle$ *for* $k \in \omega$ *with* $\sum_k 2^{-n_k} \leq 1$. *Then we can (primitive recursively) build a prefix-free machine* $M$ *and a collection of strings* $\tau_k$ *with* $|\tau_k| = n_k$ *and* $M(\tau_k) = \sigma_k$.

It is an interesting exercise to see how to use the Kraft inequality in, for example, the proof of the Calude et al. result. For instance, if $\alpha$ is c.e. then it can be constructed as a computable sequence of dyadic rationals $\alpha_s$ so that $\alpha_{s+1} - \alpha_s$ is of the form $\sum_{j \in B_s} 2^{-j}$ and we can have that $\sigma \in B_s$ implies that $\sigma$ has length less than or equal to $s$. Thus $\alpha_{s+1} - \alpha_s = p(s)2^{-(s+1)}$. Hence by Kraft's inequality we can find an $A_s$ of strings of length $s+1$ with $\alpha_{s+1} - \alpha_s = \sum_{\sigma \in A_s} 2^{-|\sigma|}$, and so that $A = \cup_s A_s$ is prefix free. (Specifically, we would enumerate $p(s)$ many requirements $\langle s+1, \lambda \rangle$.) *The beauty of the Kraft inequality is that we need only make sure that the lengths work and then the prefix-free set is implicitly given without any calculation being necessary.*

## §3. Other classes of reals.

One interesting and basically unexplored class of reals is the class of d.c.e. reals. These are defined, perhaps unfortunately, as those reals $\alpha$ for which there exist c.e. reals $\beta$ and $\gamma$ such that $\alpha = \beta - \gamma$. They are interesting since the class of c.e. reals is certainly not closed under operations such as difference. Perhaps slightly surprisingly, the d.c.e. reals form a field.

LEMMA 4 (Ambos-Spies, Weihrauch, Zheng [3]). *$\alpha$ is d.c.e. iff there exists a constant $M$ and a computable sequence of rationals $q_n$ with limit $\alpha$ such that*

$$\sum_{j=0}^{\infty} |q_{j+1} - q_j| < M.$$

PROOF. (only if) Let $x$ be d.c.e. and $x = y - z$ with $y, z$ c.e. reals. Let $y = \lim_s y_s$, and $z = \lim_s z_s$, and put $x_s = y_s - z_s$. Then $\sum_{n=0}^{\infty} |x_{n+1} - x_n| = \sum_{n=0}^{\infty} |(y_{n+1} - z_{n+1}) - (y_n - z_n)| \leq \sum_{n=0}^{\infty} |y_{n+1} - y_n| + \sum_{n=0}^{\infty} |z_{n+1} - z_n| \leq (y - y_0) + z - z_0$.

(if) Let $\sum_{n=0}^{\infty} |x_{n+1} - x_n|$ be bounded, then define $y, z$ as limits as follows.

$$y_n = x_0 + \sum_{i=0}^{n} (x_{i+1} - x_i) ; z_n = \sum_{i=0}^{n} (x_i - x_{i+1}).$$

Then the limits exist because of the bounds on the sums and one can readily verify that $x = y - z$.                                                                ⊣

QUESTION 1. *Characterize the computable g such that $\alpha$ is d.c.e. iff $\alpha(i) = \lim_s g(i, s)$ in the sense of (iv) of the Calude et al. theorem above.*

THEOREM 5 (Ambos-Spies et al. [3]). *The d.c.e. reals form a field.*

PROOF. Rearranging shows closure under addition and subtraction and multiplication. (E.g., $(x - y)(p - q) = (xp + yz) - (yp + xz)$.) Division: suppose that $x_n \to x, y_n \to y$ with $x, y$ d.c.e. and that $\sum_{n=0}^{\infty} |x_{n+1} - x_n|$, $\sum_{n=0}^{\infty} |y_{n+1} - y_n|, |x_n|, |y_n|, \frac{1}{y_n} < M$. Now

$$\sum_{n=0}^{\infty} \left| \frac{x_{n+1}}{y_{n+1}} - \frac{x_n}{y_n} \right| = \sum_{n=0}^{\infty} \left| \frac{y_n x_{n+1} - y_{n+1} x_n}{y_n y_{n+1}} \right|$$

$$\leq \sum_{n=0}^{\infty} \left| \frac{y_n x_{n+1} - y_n x_n + y_n x_n - y_{n+1} x_n}{y_n y_{n+1}} \right| \leq 2M^4.$$                ⊣

QUESTION 2. *Say something else about this field. For instance, what degrees do you get? Also what about its analytic properties such as real closure? Finally, what about its randomness properties?*

One could also ask: What about other classes of reals? For instance, we have seen that if we have a monotone increasing computable sequence of reals we get a c.e. real. What happens if we weaken the condition that the sequence be monotone as reals? As we have seen if the jumps are bounded, then we get d.c.e. reals. In general, we get the following.

THEOREM 6 (Ho [34]). *A real $\alpha$ is of the form .A for A a $\Delta_2^0$ set iff $\alpha$ is the limit of a computable set of rationals.*

PROOF. This uses another padding+density argument, as in the Calude et al. result, and is left as an exercise.                                                        ⊣

We remark that it is not difficult to show that there are d.c.e. reals that are not c.e. Here is a proof. Notice that if $D$ is a d.c.e. set (that is, $D = A - B$ for c.e. sets $A$ and $B$) then .D is a d.c.e. real.

THEOREM 7 (Ambos-Spies et al.). *There is a d.c.e. set B such that .B is neither a left nor a right computable real.*

PROOF. Let $C$ and $D$ be c.e. Turing incomparable sets. Define the d.c. set $B$ as follows.

$$B = \{4n : n \in \overline{C}\} \cup \{4n + 1 : n \in C\}$$
$$\cup \{4n + 2 : n \in D\} \cup \{4n + 3 : n \in \overline{D}\}.$$

Using the Calude et al. characterization of c.e. reals, because of the $4x, 4x + 1$ part, $\cdot\chi_B$ cannot be left computable, lest $C \leq_T D$, and similarly by the obvious modification to part $(va)$ above (reversing), the same shows that $\cdot\chi_B$ is not right computable lest $D \leq_T C$. For instance, if $\cdot\chi_B$ is left computable, let $f$ be the strongly $\Delta_2^0$ approximation given in $(v)$ of Theorem 2. Note that we can run the approximations to $C$ and $D$ and $f$ so that at each stage we can have things looking correctly. That is, we can speed the enumeration so that for all $s$ and all $n \leq s$, $n \in C_s$ iff $f(4n + 1, s) = 1$ and $f(4n, s) = 0$, $n \notin C_s$ iff $f(4n, s) = 1$ and $f(4n + 1, s) = 0$, and similarly for $D_s$, since this must be true for $C$, $D$, and $f$ in the limit. Assume that we have such enumerations. We claim that $C \leq_T D$ contrary to hypothesis. Suppose inductively we have computed $C$ up to $n - 1$. Let $s > n$ be a stage where the current approximation $f(i, s) : i \leq 4n + 3$ correctly computes $C \upharpoonright n - 1$. Note this is okay by the induction hypothesis, and means that

$$C_s \upharpoonright n - 1 = C \upharpoonright n - 1, D \upharpoonright n = D_s \upharpoonright n$$

using the $D$-oracle, and as above, $f$ appears correct for $C_s$ and $D_s$ up to $n$. Then it can only be that $n \in C$ iff $n \in C_s$. (The point is that if $n$ later enters $C$ at $t > s$ then since $f(4n, t)$ must become 0 something must enter $B$ smaller than $4n + 1$. But this is impossible since we have $C \upharpoonright n - 1 = C_s \upharpoonright n - 1$, and $D \upharpoonright n = D_s \upharpoonright n$.) The nonright computability is entirely analogous. ⊣

## §4. Degree-theoretical aspects of representations.

DEFINITION 2. *We say that a c.e. sequence of rationals $\{q_i : i \in \omega\}$ with monotonic limit $\alpha$ represents $\alpha$.*

Representations were first effectively analyzed by Calude, Coles, Hertling, and Khoussainov [7]. We have seen that if a real is c.e. then it has a computable representation. If a real is computable then every representation must be computable (exercise). Suppose that a c.e. real is noncomputable. What else can be said about its representations? For instance, the natural degree of a c.e. real is the degree of its left cut: $\deg L(\alpha)$. Does $\alpha$ always have a representation of degree $\deg L(\alpha)$? of other degrees?

THEOREM 8.

(i) (Calude, Coles, Hertling, Khoussainov) $\alpha$ *has a representation of degree* $\deg L(\alpha)$.

(ii) (Soare [58]) *If $B = \{q_i\}$ is a representation of $\alpha$ then $B \leq_T L(\alpha)$ and in fact $B \leq_{wtt} L(\alpha)$ where wtt denotes weak truth table reducibility.*[1]

(iii) (Calude et al.) *Every representation of $\alpha$ is half of a c.e. splitting of $L(\alpha)$.*

The theorem above extends earlier work of Soare who examined, in particular, the relationship between $L(a)$ and $\deg(B)$ for $a = \sum_{n \in B} 2^{-n}$. In [58], Soare observed that $L(a) \leq_T B$ and $B \leq_{tt} L(a)$. However, he also proved that there are strongly c.e. $a$, as above, with $L(a) \not\leq_{tt} B$.

Evidently (iii) implies (ii). Clearly, if $A$ represents $\alpha$ then $A$ must be an infinite c.e. subset of $L(\alpha)$. The thing to note is that $L(\alpha) - A$ is also c.e. Given rational $q$, if $q$ occurs in $L(\alpha)$, we need only wait until either $q$ occurs in $A$ or some rational bigger than $q$ does.

Note that this means that if $\alpha$ is computable then every representation of $\alpha$ is computable. Also note that the proof actually gives that if $A$ represents $\alpha$, $A \leq_{wtt} L(\alpha)$. (It is interesting to note that strong reducibilities often play a large role in effective mathematics since reducibilities that occur naturally tend to be stronger than $\leq_T$. For instance in a finitely presented group, the word problem $tt$-reduces to the conjugacy problem ([36]), algebraic closure is related to $Q$-reducibility ([5, 30, 50, 69]), and $wtt$-degrees characterize the degrees of bases of a c.e. vector space (Downey-Remmel [31]).)

We would like to prove that if $A$ is half of a splitting of $L(\alpha)$ then $A$ represents $\alpha$. But it is not difficult to prove that this is not true. We know that if $A$ represents $\alpha$ then there needs to be a computable function $g$ with range $A$ so that, as reals, $g(i) < g(i + 1)$. It is easy to construct splittings of some $\alpha$ where no such $g$ exists by a simple diagonalization argument. Calude et al. did find that the converse of (iii) did happen in some cases.

THEOREM 9 (Calude et al. [7]). *Let $A$ be a representation of $\alpha$. For subsets $B$ of $A$, the following are equivalent.*

   *$B$ represents $\alpha$.*

   *$B$ is half of a splitting of $A$.*

The proof of this result is straightforward and is left to the reader. Calude et al. [7] also obtained a partial degree theoretical converse to (iii). Namely, they showed that (i) $\alpha$ *has a representation of degree* $\deg(L(\alpha))$, *and* (ii) *every representation can be extended to one of degree* $\deg(L(\alpha))$. In [12], Downey improved the Calude et al. [7] result and obtained a complete characterization of the representations of a real $x$ in terms of the $m$-degrees of splittings of $L(x)$.

THEOREM 10 (Downey). *The following are equivalent:*

   *$b$ is the m-degree of a splitting of $L(x)$;*

   *$b$ is the m-degree of a representation of $x$.*

---

[1]We say that $A \leq_{wtt} C$ iff there is a Turing procedure $\Gamma$ and a computable function $\gamma$ such that for all $x$,
$$\Gamma(C; x) = A(x), \text{ and } u(\Gamma(C; x)) \leq \gamma(x).$$

PROOF. To prove Theorem 10, we need only show that if $L(x) = C \sqcup D$ is any c.e. splitting of $L(x)$ then there is a representation $\widehat{C} = \{c_i\}$ of $x$ of $m$-degree that of $C$. (Without loss of generality, we suppose that $C$ is noncomputable.) We do this in stages. At each stage $s$, we assume that we have enumerated $C_s$ and $D_s$ so that $L(a)_s = C_s \sqcup D_s$, where $L(a)_s$ is the collection of rationals in $L(a)$ by stage $s$, including all those of Gödel number $\leq s$. Additionally we will have a parameter $m(s)$. At stage $s + 1$ compute $C_{s+1}$ and $D_{s+1}$. Find the least rational, $q \in C_{s+1}$, by Gödel number, if any, such that $q > m(s)$.

If no such $q$ exists, set $m(s + 1) = m(s)$ and do nothing else.

If one exists, put all rationals with Gödel number below $s + 1$, in increasing real order, into $\widehat{C}_{s+1}$ and reset $m(s + 1)$ to be the maximum rational (as a real) in $L(x)_{s+1}$.

To verify the construction, first note that $\widehat{C}$ is an increasing sequence of rationals. Its limit will be $a$ provided that it is infinite, because of the use of $m(s)$.

First we claim that $m(s) \to \infty$. Suppose not, so that there is an $s$ such that, for all $t \geq s$, $m(s) = m(t)$. Then we claim that $C$ is computable, this being a contradiction. To decide if $z \in C$, go first to stage $s' = s + g(z)$, where $g(z)$ denotes the Gödel number of $z$. If $z \notin C_{s'}$, then either $z > m(s)$, or $z \in D_{s'}$. In either case, $z \notin C$. Hence $m(s) \to \infty$.

Note that $\widehat{C} \leq_m C$. Only numbers entering $C$ enter $\widehat{C}$ and can do so only at the same stage. Given $q$ go to a stage $s$ bigger than the Gödel number of $q$. If $q$ is below $m(s)$ then, as before, we can decide computably if $q \in C$. Else, note that $q \in C$ iff $q \in \widehat{C}$. The same argument shows that $C \leq_m \widehat{C}$. ⊣

We remark that many of the theorems of Calude et al. [7] now come out as corollaries to the characterization above, and known results on splittings and *wtt* degrees. Notice that by Sacks splitting theorem every noncomputable c.e. real $x$ has representations in infinitely many degrees. From known theorems we get the following.

COROLLARY 11. *There exist computably enumerable reals $a_i$ such that the collection of $T$-degrees of representations $R(a_i)$ have the following properties.*

(i) $R(a_1)$ *consists of every c.e. ($m$-)degree.*
(ii) $R(a_2)$ *forms an atomless Boolean algebra which is nowhere dense in the c.e. degrees.*

For the proofs see Downey and Stob [28].

We also remark that the above has a number of other consequences regarding known limits to splittings.

COROLLARY 12. *If a c.e. real $a$ has representations in each $T$-degree below that of $L(a)$ then either $L(a)$ is Turing complete or low$_2$.*

This follows since Downey [13] demonstrated that a c.e. degree contains a set with splittings in each c.e. degree below it iff it was complete or $\text{low}_2$. It is not clear if every nonzero c.e. degree contains a c.e. real that cannot be represented in every c.e. degree below that of $L(\alpha)$.

§5. **Presentations of reals.** The Calude et al. theorem gave many possible ways of representing reals, not just with Cauchy sequences. We explore the other methods with the following definition.

DEFINITION 3. *Let $A \subset \{0,1\}^*$. We say that $A$ is a presentation of a c.e. real $x$ if $A$ is a prefix-free c.e. set with*

$$x = \sum_{n \in A} 2^{-|n|}.$$

Previously we have seen that $x$ has representations of degree $L(x)$. However, presentations can behave quite differently.

THEOREM 13 (Downey and LaForte [27]). *There is a c.e. real $\alpha$ which is not computable, but such that if $A$ presents $\alpha$ then $A$ is computable.*

PROOF. We briefly sketch the proof, details being found in Downey and LaForte [27]. We must meet the requirements below.

$$R_e: W_e \text{ presents } \alpha \text{ implies } W_e \text{ computable.}$$

We build a computable presentation $\sum_{\sigma \in A} 2^{-|\sigma|}$ of $\alpha$, via the nearly c.e. definition. That is, we have an approximation $\alpha = \cdot a_{0,s} \ldots$ and obey the conditions that if $a_{i,s} = 1$ and $a_{i,s+1} = 0$ then $a_{j,s+1}$ becomes 1 for some $j < i$. To make $\alpha$ noncomputable, we must also meet the requirements:

$$P_e: \text{For some } i, i \in W_e \text{ iff } a_i = 1.$$

(Thus $\cdot \overline{W_e} \neq \alpha$.) The strategy for $P_e$ is simple. We must pick some $i$ to follow it and initially make it 0. At some stage $s$, if we see $i$ enter $W_e$, then we must make $a_{i,t} = 1$ for some $t \geq s$.

To make this cohere with the $R_e$, we need a little work. First, we need to surround $i$ with some 0s so that there is little interference from the other requirements, modulo finite injury. However, more importantly, we need to also make sure that for those $R_k$ of higher priority if $W_k$ presents $\alpha$ then $W_k$ is computable.

Associated with $R_k$ will be a current "length of agreement".

$$\ell(k,s) = \max\left\{ n : \alpha_s - \sum_{\sigma \in W_{k,s}} 2^{-|\sigma|} > 2^{-n} \right\},$$

We can assume that $\alpha_s > \sum_{\sigma \in W_{k,s}} 2^{-|\sigma|}$ since if a stage $t$ occurs where this is not true, we would have $\alpha_t - \sum_{\sigma \in W_{k,t}} 2^{-|\sigma|} > 2^{-d}$ for some $d$, and simply win by keeping $\alpha_s - \alpha_t < 2^{-(d+2)}$ for all stages $s < t$.

We promise that once $\ell(k, s) > d$, then no number of length $\leq d$ can enter $W_k$.

Now the idea is that when we see some $P_e$ require attention for $e$ bigger than $k$, if $i$ is smaller than $\ell(k, s)$ (the interesting case), then we wish to put a relatively big number into $a$, by changing position $i$ for the sake of $P_e$, yet we wish to not allow numbers of low length to enter $W_k$.

The idea is to slowly work backward. So first we will make position $\ell(k, s) + 1 = 1$ by adding something of length $2^{-(\ell(k,s)+1)}$ into $A_{s+1}$.

We then do nothing until $W_k$ responds by giving us a stage $t > s$ with $\ell(k, t) \geq \ell(k, s)$.

Note that $W_k$ can only change on strings of long length since we only changed $A$ slightly. Now we repeat, adding another string of the same length $2^{-(\ell(k,s)+1)}$ into $A_{t+1}$. Again we wait for another expansion stage. Note that this next addition changes things at position $\ell(k, s)$ or earlier. We can continue in this way at most $2^{\ell(k,s)-i}$ many times until we get to change position $i$. Note that we will—by restraining $\ell(k, s)$ from growing—temporarily refrain from declaring that we know $W_k$ for strings of length above $\ell(k, s)$ until a stage $t$ is found where we win $P_e$. This delay is fine since if $W_k$ actually presents $\alpha$, we will eventually get enough recovery stages that we will meet the $P_e$.

The reader should think of this as a cautious investor wishing to sell off some shares, but not allowing the market to realize this, so they drip feed the shares into the market each time the price recovers.

Now there are two outcomes. Either at some stage, we don't get recovery, so that $W_k$ does not present $\alpha$, or $W_k$ responds at each stage and we get a change only on long strings. This means that we can compute $W_k$.

Now to deal with more than one $R_k$, say $R_k$ and $R_j$, with $j < k$ we must use a tree of strategies argument. The strategy for $R_k$ guessing that $\ell(j, s) \not\to \infty$ will be to believe that the current $\ell(j, s)$ is its limit. Thus it believes that $W_j$ will not present $\alpha$ and cannot recover to its previous best length of agreement. This version of $R_k$ acts as in the basic module, but $P_e$ working with this version of $R_j$ must ensure that the total amount they could add to $\alpha - \alpha_s$ is less than $2^{-(\ell(j,s)+2)}$.

The version of $R_k$ guessing that $\ell(j, s) \to \infty$ needs to nest its expansion stages in $R_j$s. The problem is that $R_j$ can recover at many stages before $R_k$ does. During such stages, we cannot delay allowing $R_j$ to increase $\ell(j, t)$, making the allowable "quanta" even smaller. For suppose that the $P_e$ below both of these versions of $R_j$ and $R_k$ wishes to add $2^{-i}$ to $\alpha_s$. We begin this process by adding some quanta $2^{-n}$ good for both $j$ and $k$ at some stage $s_0$.

At some stage $s_1$ we might see $R_j$ recovery but not have had $R_k$ recovery. We cannot add another $2^{-n}$ to $\alpha_{s_1}$ until we get this $R_k$ recovery. On the other hand this recovery might not happen. Hence we cannot delay the extension of the definition of $W_j$ to wait for this recovery. Thus we will allow $\ell(j, s)$ to increase, lowering the quanta allowable by $R_j$. Now at stage $s_2$, after perhaps many $j$-expansionary stages we also get $R_k$ recovery. At this stage, $R_k$ would allow us to put in $2^{-n}$ but now $R_j$ only allows $2^{-m}$ for some $m >> n$.

The solution is that we won't allow another $R_k$ expansionary stage until we have had enough $j$-expansionary stages that we could (using increments of $2^{-m}$) put in $2^{-n}$. During this time we will not allow the definition of $W_j$ to be extended.

The depth $d$ strategies are similar. There are $d$ strategies $S_1, \ldots, S_d$ in decreasing order of priority. At some stage we wish to change $\alpha$ while co-operating with these $d$ strategies. We put some small quanta in. While we are waiting for recovery of *all* the $d$ strategies, we would allow the definitions of their $W_{j_d}$ to change, so for $S_1, \ldots, S_{d-1}$, the allowable quanta will be reduced. At recovery, we might wish to put $2^{-n}$ into $\alpha$ but this must now be put in quanta which is acceptable to $S_1, \ldots, S_{d-1}$. While we are doing this we delay any further work on $S_d$ until this is fulfilled. Then, like the tower of Hanoi, the same problem propagates upward. But the whole process is well-founded so that eventually progress is made. Further details can be found in Downey-Laforte [27].                                                                      ⊣

We remark that Downey and Laforte demonstrated that degrees containing such "only computably presentable" reals can be high. But if a degree is promptly simple then every c.e. real of that degree must have a noncomputable c.e. presentation. Using a $0'''$ argument, Wu [66] has constructed a c.e. noncomputable degree $\mathbf{a} \neq \mathbf{0}$ such that, if $\alpha$ is any c.e. noncomputable real of degree below $\mathbf{a}$ then $\alpha$ has a noncomputable presentation.

As with many structures of computable algebra and the like, the classification of the degrees realized as presentation seems to depend on a stronger reducibility than $\leq_T$. In this case, the relevant reducibility seems to be weak truth table reducibility.

The following is easy.

THEOREM 14. *Let $\alpha$ be a computably enumerable real, with $\alpha = .\chi_A$ for some set $A$. Suppose that $B$ is any presentation of $\alpha$. Then $B \leq_{wtt} A$ with use function the identity.*

The proof is left as an exercise. What is interesting is that there is a sort of converse to this result.

THEOREM 15 (Downey and Laforte [27]). *If $A$ is a presentation of a c.e. real $\alpha$ and $C \leq_{wtt} A$ is computably enumerable, then there is a presentation $B$ of $\alpha$ with $B \equiv_{wtt} C$.*

PROOF. Suppose $\Gamma(X)$ is a computable functional with a computable use function $\gamma$ such that $\Gamma(A) = C$. We can assume $\gamma$ is monotonically increasing. Let $\langle n, m \rangle : \mathbb{N} \times \mathbb{N} \to \mathbb{N}$ be a computable one-to-one function such that for all $n, m, \max\{n, m\} < \langle n, m \rangle$. (Adding 1 to the usual pairing function gives such a function.) Notice that, since $A$ presents $\alpha$, using the Kraft-Chaitin theorem we can enumerate strings of any length we wish into $B[s]$ as long as we ensure

$$\sum_{\sigma \in B[s]} 2^{-|\sigma|} \leq \sum_{\sigma \in A[s]} 2^{-|\sigma|}.$$

We fix enumerations of $\Gamma$, $C$, and $A$ so that at each stage $s$, exactly one element enters $C$ and for every $x < s$, $\Gamma_s(A_s; x) = C_s(x)$. We may assume $A$ is infinite, since there is nothing to prove if $A$ is computable. We construct $B$ in stages, using the function $\langle n, m \rangle$ as follows.

At stage 0, let $B[0] = \emptyset$.

At stage $s + 1$, we first find the unique number $n_s$ entering $C$ and all strings $\sigma$ that enter $A$ at stage $s + 1$. For each $|\sigma| < \gamma(n_s)$, we enumerate $2^{\langle |\sigma|, n_s \rangle - |\sigma|}$ strings of length $\langle |\sigma|, n_s \rangle$ into $B[s + 1]$. For each $|\sigma| \geq \gamma(n_s)$, we enumerate $2^{\langle |\sigma|, |\sigma|+s \rangle - |\sigma|}$ strings of length $\langle |\sigma|, |\sigma| + s \rangle$ into $B[s]$.

This ends the construction of $B$.

Notice that all of the actions taken at stage $s + 1$ serve to ensure that

$$\sum_{\sigma \in B[s+1]} 2^{-|\sigma|} = \sum_{\sigma \in A[s+1]} 2^{-|\sigma|};$$

hence, we always have enough strings available to keep $B$ prefix free.

Suppose $n \in \mathbb{N}$. Let $s(n)$ be least so that $B[s(n)]$ agrees with $B$ on all strings less than or equal to length $\langle \gamma(n), n \rangle$. Now, suppose there exists $t > s(n)$ such that $n \in C[t] - C[t - 1]$. In this case, because for every $s$ and $x < s$, $C(x)[s] = \Gamma(A; x)[s]$, there must be some $\sigma$ with $|\sigma| < \gamma(n)$ which enters $A$ at $t$. By construction, then, since $n = n_t$, we have $2^{\langle |\sigma|, n_t \rangle - |\sigma|} > 1$ strings of length $\langle |\sigma|, n_t \rangle$ entering $B$ at stage $t > s(n)$, which is a contradiction. Hence we can compute $C(n)$ from $B(n)$ with a use bounded by the number of strings of length less than or equal to $\langle \gamma(n), n \rangle$, which is a computable function. This gives $C \leq_{wtt} B$.

Next consider any binary string $\tau$. Using the computability of $\langle i, n \rangle$ and the fact that $\max\{i, n\} < \langle i, n \rangle$ we can ask whether there exist $i$ and $n$ such that $|\tau| = \langle i, n \rangle$. If not, then $\tau \notin B$. In this case, let $t(n) = 0$. Otherwise, suppose $|\tau| = \langle i, n \rangle$. If $i \geq \gamma(n)$, then $\tau$ can only enter $B$ at stage $s$ if $s = n - i$. If, on the other hand, $i < \gamma(n)$. Then if $\tau$ enters $B$ at stage $s + 1$, this can only be because $|\tau| = \langle |\sigma|, n_s \rangle$ for some $\sigma$ entering $A$ at $s$, and we enumerate $2^{\langle |\sigma|, n_s \rangle - |\sigma|}$ strings of length $\langle |\sigma|, n_s \rangle$ into $B[s + 1]$. In either case, if we let $t(n)$ be the least number greater than $n - i$ so that $C[t(n)]_{n+1} = C_{n+1}$, we have $B(\tau) = B(\tau)[t(n)]$. Since $n$ is computable from $|\tau|$, $B \leq_{wtt} C$, as required. ⊣

Note that one corollary is that a strongly c.e. real $\alpha = .A$, with the degree of $A$ $wtt$-topped,[2] has the property that it has presentations in every $T$ degree below that of $A$.

---

[2]That is, for all c.e. $B \leq_T A$, $B \leq_{wtt} A$.

Note the following.

LEMMA 16. *Suppose that A and B present $\alpha$. Then there is a presentation of $\alpha$ of wtt degree $A \oplus B$.*

The proof is to note that

$$C = \{0\sigma : \sigma \in A\} \cup \{1\sigma : \sigma \in B\}$$

is prefix free, as both $A$ and $B$ are, and presents $\alpha$.

It follows that the *wtt*-degrees of c.e. sets presenting $\alpha$ forms a $\sum_3^0$ ideal. Recently, the question of which $\sum_3^0$ ideals can represent c.e. reals was investigated by Downey and Terwijn. They combined the drip feed strategy of Downey-Laforte, a coding technique, and approximation techniques to prove the following.

THEOREM 17 (Downey and Terwijn [29]). *Suppose that $\mathcal{I}$ is any $\sum_3^0$ ideal in the computably enumerable wtt degrees. Then there is a c.e. real $\alpha$ whose degrees of presentations are exactly the members of $\mathcal{I}$.*

Downey and Terwijn also proved a sort of Rice's theorem for the index sets. Note that if $\alpha$ is a c.e. real which has a *wtt*-complete presentation, then for any $e$, $\alpha$ has a presentation of the same *wtt* degree as $W_e$. Thus the index set $\mathcal{I}(\alpha) = \{e : W_e$ has the same *wtt* degree as a presentation of $\alpha\}$ is simply $\omega$. Downey and Terwijn showed that this is the only case where this can happen. They showed that if $\alpha$ is not *wtt*-complete, then the indices in the set $\mathcal{I}(\alpha)$ are $\sum_3^0$ complete.

§6. Kolmogorov complexity. This is a theory of randomness for finite strings. In Section 8 we will look at the infinite case. The main idea is that a string is random if it is "incompressible". That is, the only way to generate the string $\tau$ is essentially to hardwire the string into the algorithm, so that the description of a program to generate $\tau$ is essentially the same size as $\tau$ itself. (For instance, $0^{10000000}$ can be described by saying that we should repeat 0 10000000 times. This can be easily described by an algorithm shorter than 10000000.)

We formalize this notion, first due to Solomonoff [60] but independently to Kolmogorov and Chaitin, as follows. Let $f : 2^{<\omega} \mapsto 2^{<\omega}$ be a partial computable function. Then we can denote the Kolmogorov complexity of a string $\sigma$ with respect to $f$ via

$$C_f(\sigma) = \min\{\infty, |p| : f(p) = \sigma\}.$$

Then relative to $f$, we say that $\sigma$ is random[3] $C_f(\sigma) \geq |\sigma|$.

---

[3]There are two traditions of notation. One is to use $K$ for Kolmogorov complexity, and $H$ for the prefix-free (disturbingly referred to as prefix complexity) complexity in the next section. The other is to use $C$ for standard Kolmogorov complexity, and to use $K$ for the prefix-free complexity of the next section. We adopt the latter and hope that it causes no confusion.

We get rid of the dependence of $f$. If we choose a universal Turing machine $U$ there is a $p$ so that $U(y, p) = f(y)$ for all $y$, then we can define $g$ via

$$g\left(0^{|p|}1py\right) = U(y, p).$$

For this $g$ we see that

$$K_g(x) \leq K_f(x) + \mathcal{O}(1).$$

Note that the constant $\mathcal{O}(1)$ is $2|p|+1$. Hence $g$ is (up to an additive constant) the minimal complexity measure.

Trivial Facts:

(1) $C(x) \leq |x| + \mathcal{O}(1)$.

(2) $C(xx) \leq C(x) + \mathcal{O}(1)$.

(3) If $h(x)$ is any total computable function, then $C(h(x)) \leq C(x) + \mathcal{O}(1)$.

We would like $C(xy) \leq C(x) + C(y) + \mathcal{O}(1)$. But this is not true. The problem is that we can't decide where $x$ finishes and $y$ starts. We get

(4) $C(xy) \leq C(x) + C(y) + 2\log|x| + \mathcal{O}(1)$.

This uses the "self-delimiting" trick used in the definition of $g$ above for the universal machine. Actually we can replace $2\log|x|$ by $2\log C(x)$.

THEOREM 18 (Solomonoff, Kolmogorov).    *For all $n$ there exists $x$ with* $|x| = n$ *and* $C(x) \geq n$.

PROOF. Count the number of strings of size $n$.                    ⊣

Notice that actually for any $k$,

$$|\{x \in 2^{<\omega} : C(x) \geq |x| - k\}| \geq 2^n \left(1 - 2^{-k}\right).$$

For instance, if $|x| = 1000$ and $k = 500$, then the number of strings of length 1000 that are "half way random" ($C(x) > 500$ is at least $2^{1000}(1 - 2^{-500})$): that is, almost every string.

Here is a simple application of the "incompressibility method."

How large is the $n$th prime? Let $m$ be a binary number and $p_i$ the largest prime divisor of $m$. To describe $m$ we need only $\langle p_i, \frac{m}{p_i}\rangle$. In fact we need only the pair $i, \frac{m}{p_i}$ ($+\mathcal{O}(1)$). Hence

$$C(m) \leq 2\log|i| + |i| + \left|\frac{m}{p_i}\right| + \mathcal{O}(1).$$

Then if $m$ is a random number, we see

$$|m| \leq 2\log|i| + |i| + \left|\frac{m}{p_i}\right| + \mathcal{O}(1).$$

Hence

$$\log m \leq 2\log\log i + \log i + \log m - \log p_i + \mathcal{O}(1).$$

Thus

$$\log p_i \leq 2 \log \log i + \log i + \mathcal{O}(1).$$

Hence $p_i \leq \mathcal{O}(i \log^2 i)$. This is pretty close to the real answer of $i \log i$.

Another classical application of Kolmogorov complexity is the construction of an immune set. Let

$$A = \left\{ x \colon C(x) \geq \frac{|x|}{2} \right\}.$$

Then $A$ is immune. Suppose that $A$ has an infinite c.e. subset $C$. Let $h(n)$ be defined as the first element of $C$ to occur in its enumeration of length above $n$. Then

$$C(h(n)) \geq |h(n)|/2 \geq n/2, \text{ but}$$
$$C(h(n)) \leq C(n) + \mathcal{O}(1) \leq |n| + \mathcal{O}(1).$$

For large enough $n$ this is a contradiction.

I should remark that there is a very well-developed theory of Kolmogorov complexity and its applications. I urge the reader to refer to Li-Vitanyi [48], especially for applications, and to refer to van Lambalgen [65] for a thorough discussion of the foundations of the subject.

§7. **Prefix-free complexity.** The main motivation for this section will be to develop a nice complexity measure for developing complexity on reals. However, Chaitin and Levin argued that prefix-free complexity (the notion we look at here) is the correct complexity even for finite strings.

Their argument for the inadequacy of classical Kolmogorov complexity is the following. The intentional meaning, it is claimed, is that the complexity of the shortest string $v$ computing $\sigma$ ought to indicate that the *bits* of $v$ containing all the information necessary to get $\sigma$ from $v$. However, they argue that the universal machine $M$ might first scan $v$ just to get its length, and only then read the bits of $v$. In this way the $v$ actually represents $|v| + \log |v|$ many bits of information. If one accepts this argument then one ought to circumvent this. Both Levin [47] and Chaitin [10] suggested special machines to circumvent this.

It has also been pointed out that one can circumvent this by asking that the complexity measure be continuous. This gives rise to the notion of *monotone complexity* which we will not deal with here. (The monotone complexity of a string $\sigma = a_1, \ldots, a_n$ is the minimum length of a string $\tau$ such that the universal machine $U$ with oracle $\tau$ and argument $m$ computes $a_1, \ldots, a_m$ for all $m \leq n$.) We refer the reader to Li-Vitanyi [48] and Bārzdiņš [4] for more details.

A prefix-free machine is one whose domain is prefix free. It is usual to take the machine as "self delimiting" which means that it has a one way read head which halts when the machine accepts and has accepted the string described by the read head up to its present position. This is a purely technical device that forces the machine to have a prefix-free domain.

Facts:

(1) If $\varphi$ is a partial computable function with prefix-free domain, then there is a prefix-free (self-delimiting) machine $M$ such that $M$ agrees with $\varphi$.

(2) There is a universal (self-delimiting) prefix-free machine.

The reader should prove these not altogether obvious results. The reason we want prefix-free machines is that we will be looking at reals (eventually) and we wish to apply Kraft's inequality. So the domains will need to be prefix free. Let $K(x)$ denote the prefix-free Kolmogorov complexity of $x$. The counting arguments of the previous section demonstrate that

$$K(x) \le |x| + 2\log|x| + \mathcal{O}(1),$$
$$K(x) \le |x| + K(|x|) + \mathcal{O}(1).$$

We remark that this is tight. A counting argument shows that there are many $x$ with $K(x) > |x| + \log|x|$. (The readers are advised to prove this for themselves.[4] In fact for *finite strings* this would be the typical notion of randomness used if we demanded prefix-free complexity.) There is one good tradeoff, namely, now $K(xy) \le K(x) + K(y) + \mathcal{O}(1)$. For these and other facts we refer the reader to Li-Vitanyi [48] or Fortnow [32].

The actual relationships between $C$ and $K$ are

$$K(x) = C(x) + C(C(x)) + \mathcal{O}(C(C(C(x)))).$$
$$C(x) = K(x) - K(K(x)) + \mathcal{O}K^{(3)}(x).$$

These are due to Solovay in the 1975 manuscript and are nontrivial.

## §8. Complexity of reals.
Our first attempt to define a random real would be to define $\alpha = .A$ to be random iff there was a constant $\mathcal{O}(1)$ such that for all $n$, $C(A \restriction n) + \mathcal{O}(1) > n$. Unfortunately *no* real satisfies this condition.

To see this, we know that for all $k$ there is an $n$ such that $n$ is a program for $A \restriction k$. Let $c$ be the fixed constant with $C(x) \le x + c$. Then

$$K(A \restriction n) \le K(A \restriction n - k) \le n - k + c.$$

Here we are explicitly using the fact that the *length $n$* of $A \restriction n$ gives additional information. This can easily be improved upon. For instance, it can be shown that $C(A \restriction n) \le n - \log n$ infinitely often. (See Li-Vitanyi [48], p. 138.)

However, all problems are removed if we use $K$ in place of $C$.

---

[4]Suppose that for all $x$, $K(x) \le |x| + \log|x|$. Then $\sum_{\sigma \in \Sigma^*} 2^{-K(\sigma)} \ge \sum_{\sigma \in \Sigma^*} 2^{-(|\sigma| + \log|\sigma|)} \ge \sum_n \sum_{|\sigma|=n} 2^{-(n + \log n)} \ge \sum_n 2^n (2^{-(n + \log n)}) \ge \sum_n 1/n = \infty$, a contradiction, since $\sum_\sigma 2^{-K(\sigma)} < 1$.

DEFINITION 4 (Chaitin-Levin). *A real* $\alpha = .A$ is Chaitin random *if there is an $\mathcal{O}(1)$ such that, for all $n$,*

$$K(A \restriction n) \geq n - \mathcal{O}(1).$$

It can be shown that if

$$K(A \restriction n) \geq n - \mathcal{O}(1) \text{ for } \textit{infinitely many } n,$$

(this being called *Kolmogorov* random) then the real $\alpha$ is Chaitin random. Unfortunately Schnorr [56] proved that the converse does not hold. (It can, however, be shown that the set of languages which are Chaitin random but not Kolmogorov random has measure zero.) There are reals that are Kolmogorov random but not Chaitin random. Before we prove the existence of such a real, we look at other (and earlier) topological views of randomness.

The main idea is that a real would be random iff it had no rare properties. Using measure theory, this translates as no "effectively null" properties. We define a c.e. open set to be a c.e. collection of open rational intervals. The first guess one might make for a random real is that

"a real $x$ is random iff for all computable collections of c.e. open sets $\{U_n : n \in \omega\}$, with $\mu(U_n) \to 0$, $x \notin \cap_n U_n$."

This is a very strong definition and is stronger than the most commonly accepted version of randomness. Let's call this *strong randomness*.[5] The key is that we wish to avoid all "effectively null" sets. Surely an effectively null set would be one where the measures went to zero in some computable way. Such considerations lead to the definition of Martin-Löf randomness below.

DEFINITION 5 (Martin-Löf, [51]). *We say that a real is* Martin-Löf random *or 1-random iff for all computable collections of c.e. open sets $\{U_n : n \in \omega\}$, with $\mu(U_n) \leq 2^{-n}$, $x \notin \cap_n U_n$.*

We call a computable collection of c.e. open sets a *test*, and ones with $\mu(U_n) \leq 2^{-n}$ for all $n$, a Martin-Löf test. The usual terminology is to say that a real is Martin-Löf random if it passes all Martin-Löf tests. Of course, a real passes the test if it is not in the intersection.

We remark that while strong randomness clearly implies Martin-Löf randomness, the converse is not true. This is an observation of Solovay. Later we show that there are c.e. reals that are Martin-Löf random. Hence the inequivalence of strong randomness and Martin-Löf randomness will follow by showing that no strong random real is c.e. The following proof of this observation is due to Martin (unpublished).

Let $\alpha = \lim_s q_s$ as usual and define

$$U_n = \left\{ y : \exists s \geq n \left[ y \in (q_n, q_n + 2(q_s - q_n)) \right] \right\}.$$

---

[5]This notion has been examined. It is equivalent to $A$ is in every $\Sigma_2^0$ class of measure 1. Kurtz and Kautz call this notion *weakly $\Sigma_2^0$-random*. It was also used by Gaifman and Snir. The reader is referred to Li-Vitanyi [48], p. 164, where they call it $\Pi_2^0$-randomness.

Then $\mu(U_n) \to 0$, yet $\alpha \in \cap_n U_n$. (Actually this shows that $\alpha$ cannot even be $\Delta_2^0$.)

In a famous unpublished manuscript, Solovay proposed an alternative notion of randomness.

DEFINITION 6 (Solovay [61]). *We say that a real $x$ is* Solovay random *iff for all computable collections of c.e. $\{U_n : n \in \omega\}$ such that $\sum_n \mu(U_n) < \infty$, $x$ is in only finitely many $U_i$.*

The reader should note the following alternative version of Definition 6.

*A real is* Solovay random *iff for all computably enumerable collections of rational intervals $I_n : n \in \omega$, if $\sum_n |I_n| < \infty$, then $x \in I_n$ for at most finitely many $n$.*

Again, we can define a Solovay test as a collection of rational intervals $\{I_i : i \in \omega\}$, with $\sum_i I_i < \infty$. Then a real is Solovay random iff it passes every Solovay test, meaning that it is in only finitely many $I_i$. Clearly if $x$ is Solovay random, then it is Martin-Löf random. The converse also holds.

THEOREM 19 (Solovay [61]). *A real $x$ is Martin-Löf random iff $x$ is Solovay random.*

PROOF. Suppose that $x$ is Martin-Löf random. Let $\{U_n\}$ be a computable collection of c.e. open sets with $\sum_n \mu(U_n) < \infty$. We can suppose, by leaving some out, that $\sum_n \mu(U_n) < 1$. Define a c.e. open set as

$$V_k = \{y \in (0,1) : y \in U_n \text{ for at least } 2^k \ U_n\}.$$

Then $\mu(V_k) \leq 2^{-k}$ and hence as $x$ is Martin-Löf random, $x \notin \cap_n V_n$, giving the result. ⊣

It is also true that Chaitin random is equivalent to Martin-Löf random.

THEOREM 20 (Schnorr). *A real $x$ is Chaitin-Levin random iff it is Martin-Löf random.*

PROOF. ($\to$) Suppose that $x$ is Martin-Löf random. Let

$$U_k = \{y : \exists n K(y \restriction n) \leq n - k\}.$$

Recall that there is a $C$ such that (for a fixed $n$),

$$\mu(\{y : H(y \restriction n) \leq K(n) + n - k\}) \leq C2^{-k},$$

and hence

$$\mu(\{y : H(y \restriction n) \leq n - k\}) \leq C2^{-K(n)-k}$$

Now we can estimate the size of $U_k$:

$$\mu(U_k) \leq C2^{-k} \left( \sum_{n=1}^{\infty} 2^{-K(n)} \right) \leq 2^{-k}.$$

Hence the sets $\{U_k : k \in \omega\}$ form a Martin-Löf test, and if $x$ is Martin-Löf random $x \notin \cap_n U_n$. Thus there is a $k$ such that, for all $n$, $K(x \restriction n) > n - k$. ⊣

The other direction of the proof is more difficult, and the most elegant proof known to the author is the one of Chaitin [10]. This approach is slightly more abstract since it is *axiomatic* and stresses the *minimality* of $K$ as a measure of complexity. It is thus of interest in its own right.

Specifically, Chaitin defined an *information content measure* as any function $\widehat{H}$ such that

$$\Omega_{\widehat{H}} = \sum_{\sigma \in 2^{<\omega}} 2^{-\widehat{H}(\sigma)} < 1 \text{ and}$$

$$\left\{ \langle \sigma, k \rangle : \widehat{H}(\sigma) \leq k \right\} \text{ is c.e.}$$

Naturally one can enumerate the information content measures $\{H_k : k \in \omega\}$[6] and then can define

$$K(x) = \min_{k \geq 0} \{H_k(x) + k + 1\}.$$

Notice that by the universal Turing machine, $\{\langle \sigma, k \rangle : H(\sigma) \leq k\}$ is c.e. Furthermore,

$$\sum_{\sigma} 2^{-K(\sigma)} = \sum_{k \geq 1} 2^{-k} \left( \sum_{\sigma} 2^{-H_k(\sigma)} \right) < 1.$$

Notice therefore that for any information content measure

$$K(\sigma) \leq H_k(\sigma) + \mathcal{O}(1).$$

Thus we see that (of course) $K$ is the prefix free Kolmogorov complexity, and this information content measure is minimal among all such measures.

Before we turn to the proof of the other direction of Schnorr's theorem, here is one application of this idea. We prove that

$$K(x) \leq |x| + K(|x|) + \mathcal{O}(1).$$

This result was mentioned before, but we only alluded to a proof suggesting that it was merely a counting argument. Here is Chaitin's proof.

$$1 > \Omega = \sum_{x \in 2^{<\omega}} 2^{-K(x)} = \sum_{n \in \mathbb{N}} \left[ 2^{-n} \sum_{|x|=n} 2^{-K(x)} \right], \text{ (that same trick)},$$

$$= \sum_{n} \sum_{|x|=n} 2^{-(n+K(x))} = \sum_{x \in 2^{<\omega}} 2^{-(|x|+K(|x|))}.$$

Thus $K(x) \leq |x| + K(|x|) + \mathcal{O}(1)$, as $K$ is minimal.

The reader should think of proofs like this as using the minimality of $K$ to avoid explicit mention of Kraft-Chaitin. Now back to the proof of Schnorr's Theorem:

---

[6]It is easy to spot when the measure threatens to exceed 1, at which point one would stop enumerating a bad $M_k$.

($\leftarrow$) This time suppose that $x$ is not Martin-Löf random. We prove that $x$ is not Chaitin random. Thus we have $\{U_n\}$ with $x \in \cap U_n$ and $\mu(U_n) \leq 2^{-n}$. We note that $\sum_n 2^{-n^2+n}$ converges, and indeed, $\sum_{n \geq 3} 2^{-n^2+n} < 1$. Notice that

$$\sum_{n \geq 3} \sum_{\sigma \in U_{n^2}} 2^{-(|\sigma|-n)} \leq \sum_{n \geq 3} 2^n \mu(U_{n^2}) \leq \sum_{n \geq 3} 2^{-n^2+n} < 1.$$

Thus by the minimality of $K$, $\sigma \in U_{n^2}$ and $n \geq 3$ implies that $K(\sigma) \leq |\sigma| - n + \mathcal{O}(1)$. Therefore, as $x \in \cap U_{n^2}$ for all $n \geq 3$ we see that $K(x \restriction k) \leq k - n + \mathcal{O}(1)$, and hence it drops arbitrarily away from $k$. Hence, $x$ is not Chaitin random.                    $\dashv$

If the reader wished to reinstate Kraft-Chaitin here, then the argument above is roughly the following. Since $x \in U_{n^2}$ (or any reasonable function of $n$, $2n$ would probably be enough), since the measure is small ($< 2^{-n^2}$), we can use Kraft-Chaitin to enumerate a machine which maps strings of length $k - n$ to initial segments of length $k$ of strings in $U_{n^2}$. Specifically, as we see strings $\sigma$ with $I(\sigma) \in U_{n^2}$ and length at least $n^2$, then we could enumerate a requirement $|\sigma| - k, \sigma$. (The total measure will be bounded by 1 and hence Kraft-Chaitin applies.)

We note that at this stage, we have not yet any examples of random reals. Here is one due to Chaitin. Fix a universal prefix-free machine $M$.

$$\Omega_M = \sum_{n(\sigma)\downarrow} 2^{-|\sigma|}.$$

Note that $\Omega$ is a c.e. real. As we see in the next section, it is random, and among c.e. reals, in some sense the *only* random real.

## §9. Relative randomness.

We wish to look at reals, especially c.e. reals under notions of relative randomness. Ultimately, we would seek to understand $\leq_K$ and $\leq_C$ reducibilities, for instance, where for $E = K$ or $C$, we have

$$\alpha \leq_E \beta \text{ iff } \forall n \left[ E(\alpha \restriction n) \leq E(\beta \restriction n) + \mathcal{O}(1) \right].$$

There are a number of natural reducibilities which imply $\leq_E$. One was introduced by Solovay and some are more recent. In this section we will look at some recent material on such reducibilities.

DEFINITION 7 (Solovay [61]). *We say that a real $\alpha$ is Solovay reducible to $\beta$ (or $\beta$ dominates $\alpha$), $\alpha \leq_S \beta$ iff there is a constant $c$ and a partial computable function $f$, so that for all $q \in \mathbb{Q}$, with $q < \beta$,*

$$c(\beta - q) > \alpha - f(q).$$

The intuition is that a sequence of rationals converging to $\beta$ can be used to generate one converging to $\alpha$ at the same rate. The point is that if we have a c.e. sequence $\{q_n : n \in \omega\}$ of rationals converging to $\beta$ then we know that $f(q_n)\downarrow$. Notice that if $r_n \to \alpha$ then for all $m$ there is some $k$ such that $\alpha > r_k > f(q_m)$.

(The reals are not rational.) Noticing this yields the following characterization of Solovay reducibility.

LEMMA 21 (Calude et al. [7]). *For c.e. reals,* $\alpha \leq_S \beta$ *iff for all c.e.* $q_i \to \beta$ *there exists a total computable g and a constant c such that, for all m,*

$$c(\beta - q_m) > \alpha - r_{g(m)}.$$

Another characterization of $\leq_S$ is the following.

THEOREM 22 (Downey, Hirschfeldt, Nies [23]). *For c.e. reals,* $\alpha \leq_S \beta$ *iff for all c.e. sequences* $\{q_i : i \in \omega\}$ *such that* $\beta = \sum_i q_i$, *there is a computable function* $: \omega \mapsto [0, 1]$ *and a constant c such that,*

$$\alpha = c \left( \sum_i (i)q_i \right).$$

*Hence* $\alpha \leq_S \beta$, *iff there exists a c and a c.e. real* $\gamma$ *such that*

$$c\beta = \alpha + \gamma.$$

PROOF. (if) One direction is easy. Suppose that $c$ and   exist. Notice that

$$c \left( \beta - \sum_{i=1}^n q_i \right) > \alpha - \sum_{i=1}^n (i)q_i.$$

Hence $\alpha \leq_S \beta$.                  ⊣

For the other direction, we need the following lemmas. The first is implicit in Solovay's manuscript but is first proven in [22].

LEMMA 23. *Let* $\alpha$ *and* $\beta$ *be c.e. reals, and let* $\alpha_0, \alpha_1, \ldots$ *and* $\beta_0, \beta_1, \ldots$ *be computable increasing sequences of rationals converging to* $\alpha$ *and* $\beta$, *respectively. Then* $\beta \leq_S \alpha$ *if and only if there are a constant d and a total computable function f such that for all* $n \in \omega$,

$$\beta - \beta_{f(n)} < d(\alpha - \alpha_n).$$

The proof is straightforward and is left as an exercise.

LEMMA 24 (Downey, Hirschfeldt, Nies [23]). *Let* $\beta \leq_S \alpha$ *be c.e. reals and let* $\alpha_0, \alpha_1, \ldots$ *be a computable increasing sequence of rationals converging to* $\alpha$. *There is a computable increasing sequence* $\hat{\beta}_0, \hat{\beta}_1, \ldots$ *of rationals converging to* $\beta$ *such that for some constant c and all* $s \in \omega$,

$$\hat{\beta}_s - \hat{\beta}_{s-1} < c(\alpha_s - \alpha_{s-1}).$$

PROOF. Fix a computable increasing sequence $\beta_0, \beta_1, \ldots$ of rationals converging to $\beta$, let $d$ and $f$ be as in Lemma 23, and let $c > d$ be such that $\beta_{f(0)} < c\alpha_0$. We may assume without loss of generality that $f$ is increasing. Define $\hat{\beta}_0 = \beta_{f(0)}$.

There must be an $s_0 > 0$ for which $\beta_{f(s_0)} - \beta_{f(0)} < d(\alpha_{s_0} - \alpha_0)$, since otherwise we would have

$$\beta - \beta_{f(0)} = \lim_s \beta_{f(s)} - \beta_{f(0)} \geq \lim_s d(\alpha_s - \alpha_0) = d(\alpha - \alpha_0),$$

contradicting our choice of $d$ and $f$. It is now easy to define $\hat{\beta}_1, \ldots, \hat{\beta}_{s_0}$ so that $\hat{\beta}_0 < \cdots < \hat{\beta}_{s_0} = \beta_{f(s_0)}$ and $\hat{\beta}_s - \hat{\beta}_{s-1} \leq d(\alpha_s - \alpha_{s-1}) < c(\alpha_s - \alpha_{s-1})$ for all $s \leq s_0$. For example, if we let $\mu$ be the minimum value of $d(\alpha_s - \alpha_{s-1})$ for $s \leq s_0$ and let $t$ be least such that $\hat{\beta}_0 + d(\alpha_t - \alpha_0) < \beta_{f(s_0)} - 2^{-t}\mu$ then we can define

$$\hat{\beta}_{s+1} = \begin{cases} \hat{\beta}_s + d(\alpha_{s+1} - \alpha_s) & \text{if } s + 1 < t \\ \beta_{f(s_0)} - 2^{-(s+1)}\mu & \text{if } t \leq s + 1 < s_0 \\ \beta_{f(s_0)} & \text{if } s + 1 = s_0. \end{cases}$$

We can repeat the procedure in the previous paragraph with $s_0$ in place of 0 to obtain an $s_1 > s_0$ and $\hat{\beta}_{s_0+1}, \ldots, \hat{\beta}_{s_1}$ such that $\hat{\beta}_{s_0} < \cdots < \hat{\beta}_{s_1} = \beta_{f(s_1)}$ and $\hat{\beta}_s - \hat{\beta}_{s-1} < c(\alpha_s - \alpha_{s-1})$ for all $s_0 < s \leq s_1$.

Proceeding by recursion in this way, we define a computable increasing sequence $\hat{\beta}_0, \hat{\beta}_1, \ldots$ of rationals with the desired properties. $\dashv$

We are now in a position to prove Theorem 22 for the other direction.

PROOF. (only if of Theorem 22) Suppose that $\beta \leq_S \alpha$. Given a computable sequence of rationals $a_0, a_1, \ldots$ such that $\alpha = \sum_{n \in \omega} a_n$, let $\alpha_n = \sum_{i \leq n} a_i$ and apply Lemma 24 to obtain $c$ and $\hat{\beta}_0, \hat{\beta}_1, \ldots$ as in that lemma. Define $\varepsilon_n = (\hat{\beta}_n - \hat{\beta}_{n-1})a_n^{-1}$. Now $\sum_{n \in \omega} \varepsilon_n a_n = \sum_{n \in \omega} \hat{\beta}_n - \hat{\beta}_{n-1} = \beta$, and for all $n \in \omega$,

$$\varepsilon_n = \left(\hat{\beta}_n - \hat{\beta}_{n-1}\right) a_n^{-1} = \left(\hat{\beta}_n - \hat{\beta}_{n-1}\right) (\alpha_n - \alpha_{n-1})^{-1} < c. \qquad \dashv$$

What has this to do with Kolmogorov complexity? The following lemma of Solovay is the decisive fact we use for this (and other) reducibilities.

LEMMA 25 (Solovay). *For all $k$ there is a constant $c_k$ depending on $k$ alone, such that for all $n$, $|\sigma| = |\tau| = n$ and $|\sigma - \tau| < 2^{k-n}$, then for $E = H$ or $C$,*

$$E(\sigma) \leq E(\tau) + c_k.$$

PROOF. Here is the argument for $C$. We can write a program depending on $k$ which, when given $\sigma$, reads the length of $\sigma$ then computes the $v$ such that $v$ has the same length as $\sigma$ and $|\sigma - v| < 2^{k-n}$. Then, given a program for $\sigma$, all we need to generate $\tau$ is to use the program for the $v$s and compute which $v$ is $\tau$ on the list. This is nonuniform, but only needs about $\log k$ many bits since the size of the list depends on $k$ alone.

The argument for $K$ is similar. Suppose that we have a prefix-free $M$. When we see some $v$ with $M(v) = \sigma$, then we can enumerate a requirement

$|v| + 2^{k+1}, \tau$ for each of the $2^k$ $\tau$ with $|\sigma - \tau| < 2^{k-n}$. Now apply Kraft-Chaitin. $\dashv$

Now we use Lemma 25 to relate Solovay reducibility to complexity.

THEOREM 26 (Solovay). *Suppose that* $\alpha \leq_S \beta$. *Then for* $E = H$ *or* $C$, $\alpha \leq_E \beta$.

PROOF. Suppose that $\alpha \leq_S \beta$ via $c < 2^k$, $f$. Notice that

$$\alpha - f\left(\beta \restriction (n+1)\right) < 2^k \left(\beta - \beta \restriction (n+1)\right).$$

In particular,

$$\alpha \restriction n - f\left(\beta \restriction (n+1)\right) \restriction n < 2^{k-n},$$

and we can apply Lemma 25. $\dashv$

It is natural to try to understand the nature of Solovay reducibility on the c.e. reals and how precisely it relates to $\leq_K$ and $\leq_C$.

First Solovay noted that $\Omega$ was Solovay complete. (Be aware that this means for all c.e. *reals* (not just c.e. *sets*) $\alpha$, $\alpha \leq_S \Omega$. This is obvious from the definition. Furthermore if $\Omega \leq_S \alpha$, for any (not necessarily c.e.) real $\alpha$ then $\alpha$ must be random. This follows by the Chaitin definition of randomness and by Theorem 26. Finally Kučera and Slaman showed that domination provides a precise characterization of randomness.

THEOREM 27 (Kučera and Slaman [43]). *Suppose that* $\alpha$ *is random and c.e. Then for all c.e. reals* $\beta$, $\beta \leq_S \alpha$.

PROOF. Suppose that $\alpha$ is random and $\beta$ is a c.e. real. We need to show that $\beta \leq_S \alpha$. We enumerate a Martin-Löf test $\{F_n : n \geq 1\}$ in stages. Let $\alpha_s \to \alpha$ and $\beta_s \to \beta$ computably and monotonically. We assume that $\beta_s < \beta_{s+1}$. There will be a parameter $t_{n,s}$ associated with the test set $F_n^s$. Initially, $t_{n,0} = 0$. At stage $s$ if $\alpha_s \in F_n^s$, do nothing, else put $(\alpha_s - 2^{-(n+1)}(\beta_{s+1} - \beta_{t_{n,s}}), \alpha_s + 2^{-(n+1)}(\beta_{s+1} - \beta_{t_{n,s}}))$ into $F_n^{s+1}$, setting $t_{m,s+1} = s + 1$, for all $m \geq n$. One verifies that $\mu(F_n) < 2^{-n}$. Thus the $F_n$ define a Martin-Löf test. As $\alpha$ is random, there is an $n$ such that for all $m \geq n$, $\alpha \notin F_m$. This shows that $\beta \leq_S \alpha$ with constant $2^n$. $\dashv$

So we see that Solovay reducibility is good with respect to randomness. Notice that Kučera and Slaman's theorem says something very strong. Consider a random c.e. real $x$. Then for $K$, e.g., we know that

$$K(x \restriction n) \geq n + \mathcal{O}(1).$$

However, we also know that

$$K(\sigma) \leq |\sigma| + 2\log(|\sigma|) + \mathcal{O}(1).$$

It would seem that there could be random $y$ and random $x$ where for infinitely many $n$, $x \restriction n$ had $K$-complexity $n + \log n$, yet $y$ had $K$-complexity $n$. Why not? After all, the complexity only needs to be above $n$ to "qualify" as random, and it certainly can be as large as $n + \log n$.

*However, Kučera and Slaman's theorem says that this is not so. All random c.e. reals have "high" complexity (like $n + \log n$) and low complexity (like $n$) at the same $n$'s!* Similarly, for $C$, a real $x$ is random iff

$$K(x \mid n) \geq n + \mathcal{O}(1)$$

infinitely often. This definition is enough to guarantee that the reals have the same $C$-complexity for all $n$, a remarkable fact.

Before we turn to the structure of the Solovay degrees of c.e. reals, we mention that we know of no characterization of this (or any of the other reducibilities we examine) in terms of *test sets*. What we are thinking here is that $\alpha \leq \beta$ iff every test failed by $\beta$ is failed by $\alpha$, or something. This, of course is not correct since $\alpha$ and $\beta$ will no doubt be different rational intervals. But there should be some computable map, perhaps from test sets to test sets, like an $m$-reduction which will be along these lines.[7]

§10. **The structure of Solovay degrees of c.e. reals.** Despite the many attractive features of the Solovay degrees of c.e. reals, their structure is largely unknown. Recently progress has been made.

THEOREM 28 (Downey, Hirschfeldt, and Nies [23]). *The Solovay degrees of c.e. reals*

(i) *form a distributive upper semilattice, where the operation of join is induced by $+$, arithmetic addition (or multiplication) (namely, $[x] \vee [y] \equiv_S [x+y]$.);*
(ii) *are dense;*
(iii) *if $a$ is incomplete and $b <_S a$, then there exist $a_1 |_S a_2$ such that $b < a_1, a_2$, and $a = a_1 \vee a_2$; that is, every incomplete degree splits over all lesser ones;*
(iv) *if $[\Omega] = a \vee b$ then either $[\Omega] = a$ or $[\Omega] = b$.*

PROOF. We will sketch the proof of some of the above. We begin with (i). We will be applying Theorem 22 but for convenience will write $_i$ instead of $(i)$.

Suppose that $\beta \leq_S \alpha_0 + \alpha_1$. Let $a_0^0, a_1^0, \ldots$ and $a_0^1, a_1^1, \ldots$ be computable sequences of rationals such that $\alpha_i = \sum_{n \in \omega} a_n^i$ for $i = 0, 1$. By Theorem 22, there is a constant $c$ and a computable sequence of rationals $\varepsilon_0, \varepsilon_1, \cdots < c$ such that $\beta = \sum_{n \in \omega} \varepsilon_n(a_n^0 + a_n^1)$. Let $\beta_i = \sum_{n \in \omega} \varepsilon_n a_n^i$. Then $\beta = \beta_0 + \beta_1$ and, again by Theorem 22, $\beta_i \leq_S \alpha_i$ for $i = 0, 1$. This establishes distributivity.

To see that the join in the Solovay degrees is given by addition, we again apply Theorem 22. Certainly, for any c.e. reals $\beta_0$ and $\beta_1$ we have $\beta_i \leq_S \beta_0 + \beta_1$ for $i = 0, 1$, and hence $[\beta_0 + \beta_1] \geq_S [\beta_0], [\beta_1]$. Conversely, suppose that $\beta_0, \beta_1 \leq \alpha$. Let $a_0, a_1, \ldots$ be a computable sequence of rationals such that $\alpha = \sum_{n \in \omega} a_n$. For each $i = 0, 1$ there is a constant $c_i$ and a computable

---

[7]Downey, Griffiths, and Hirschfeldt [17] have developed a notion of *submartingale reducibility* that works along these lines, and may well have applications in subcomputable classes.

sequence of rationals $\varepsilon_0^i, \varepsilon_1^i, \cdots < c_i$ such that $\beta_i = \sum_{n \in \omega} \varepsilon_n^i a_n$. Thus $\beta_0 + \beta_1 = \sum_{n \in \omega} (\varepsilon_n^0 + \varepsilon_n^1) a_n$. Since each $\varepsilon_n^0 + \varepsilon_n^1$ is less than $c_0 + c_1$, a final application of Theorem 22 shows that $\beta_0 + \beta_1 \leq_S \alpha$. Multiplication is similar.

Now we turn to the density properties. The proof that if $\alpha <_S \Omega$ then there is a $\beta$ with $\alpha <_S \beta <_S \Omega$ is a relatively straightforward finite injury argument, based especially on the fact that we don't need to actually build the reduction $\beta \leq \Omega$. We omit this proof. The argument that every incomplete degree splits over all lesser ones has some novel features.

We will prove the following.

Let $\gamma <_S \alpha <_S \Omega$. There are $\beta^0$ and $\beta^1$ such that $\gamma <_S \beta^0, \beta^1 <_S \alpha$ and $\beta^0 + \beta^1 = \alpha$.

Recall that $\alpha \leq_S \beta$ iff there is a computable $f$ and a constant $d$ such that $\alpha - \alpha_{f(n)} < d(\beta - \beta_n)$ for all $n$.

We want to build $\beta^0$ and $\beta^1$ such that

(i) $\beta^0, \beta^1 \leq_S \alpha$,
(ii) $\beta^0 + \beta^1 = \alpha$, and
(iii) the following requirement is satisfied for each $e, k \in \omega$ and $i < 2$:

$$R_{i,e,k}: \Phi_e \text{ total} \Rightarrow \exists n \left( \alpha - \alpha_{\Phi_e(n)} \geq k \left( \beta^i - \beta_n^i \right) \right).$$

The argument is finite injury; however, there are several problems with the implementation. It suffices to discuss a two-requirement scenario.

$R_0: \Phi \text{ total} \Rightarrow \exists n(\alpha - \alpha_{\Phi(n)} \geq k(\beta^0 - \beta_n^0))$
$R_1: \Psi \text{ total} \Rightarrow \exists n(\alpha - \alpha_{\Psi(n)} \geq l(\beta^1 - \beta_n^1))$

Naturally, we will be measuring whether $\Phi$ and $\Psi$ are total and only work when this appears so. Thus, without loss of generality, we will assume that $\Phi$ and $\Psi$ are total. The reader should imagine the construction as follows. There are

two containers labeled $\beta^0$ and $\beta^1$, and
a large funnel through which bits of $\alpha$ are being poured.

As with any priority argument, $R_0$ and $R_1$ fight for control of the funnel. In particular, bits of $\alpha$ must go into the containers (because we want $\beta^0 + \beta^1 = \alpha$) at the same rate as they go into $\alpha$ (because we want $\beta^0, \beta^1 \leq_S \alpha$). However, each $R_i$ wants to funnel enough of $\alpha$ into $\beta^{1-i}$ to be satisfied.

As $R_0$ is stronger, it could potentially put all of $\alpha$ into $\beta^1$, but that would leave $R_1$ unsatisfied. The trouble comes from trying to recognize when enough of $\alpha$ has been put into $\beta^1$ so that $R_0$ is satisfied.

DEFINITION 8. $R_0$ is satisfied through $n$ at stage $s$ if $\Phi(n)[s]\downarrow$ and $\alpha_s - \alpha_{\Phi(n)} > k(\beta_s^0 - \beta_n^0)$.

To achieve satisfaction, the idea is that $R_0$ sets a quota for $R_1$ (how much may be funneled into $\beta^0$ from that point on). If the quota is $2^{-m}$ and $R_0$ finds that either

(i) it is unsatisfied or

(ii) the least number through which it is satisfied changes,

then it sets a new quota of $2^{-(m+1)}$ for how much may be funneled into $\beta^0$ from that point on.

LEMMA 29. *There is an $n$ through which $R_0$ is eventually permanently satisfied, that is,*

$$\exists n, s \, \forall t > s \, \left(\alpha_t - \alpha_{\Phi(n)} > k \left(\beta_t^0 - \beta_n^0\right)\right).$$

PROOF OF LEMMA. Suppose not. Then $R_1$s quota $\to 0$, so $\beta^0$ is computable. Also, $\forall n, s \, \exists t > s \, [\alpha_t - \alpha_{\Phi(n)} \le k(\beta_t^0 - \beta_n^0)]$. So

$$\forall n \, [\alpha - \alpha_{\Phi(n)} < (k+1)(\beta^0 - \beta_n^0)].$$

Thus $\alpha \le_S \beta^0$ is computable. Contradiction.                       $\dashv$

Thus the strategy above yields a method for meeting $R_0$. At the end of this process, $R_0$ is permanently satisfied, and $R_1$ has a final quota $2^{-m}$ that it is allowed to put into $\beta^0$.

Now we hit the crucial problem, precisely where we need incompleteness for $\alpha$. If $R_1$ waits until a stage $s$ such that $\alpha - \alpha_s < 2^{-m}$ then it can put all of $\alpha - \alpha_s$ into $\beta^0$ and $t$ will, in turn, be satisfied.

The problem is that $R_1$ *cannot tell when such an $s$ arrives.* If $R_1$ jumps the gun, it may find itself unsatisfied and unable to do anything about it since it will have used all of its quota *before* $s$ arrives.

The key new idea is that $R_1$ *uses $\Omega$ as an investment advisor.*

Let $s$ be the stage at which $R_1$s final quota of $2^{-m}$ is set. At each stage $t \ge s$, $R_1$ puts as much of $\alpha_{t+1} - \alpha_t$ into $\beta^0$ as possible so that the total amount put into $\beta^0$ since stage $s$ *does not exceed* $2^{-m}\Omega_t$. The total amount put into $\beta^0$ after stage $s$ is $\le 2^{-m}\Omega < 2^{-m}$, so the quota is respected. We finish the proof with the following lemma.

LEMMA 30. *There is a stage $t$ after which $R_1$ is allowed to funnel all of $\alpha - \alpha_t$ into $\beta^0$.*

PROOF. It is enough that $\exists u \ge t \ge s \, \forall v > u \, (2^{-m}(\Omega_v - \Omega_t) \ge \alpha_v - \alpha_t)$. Suppose not. Then $\forall u \ge t \ge s \, \exists v > u \, [\Omega_v - \Omega_t < 2^m(\alpha_v - \alpha_t)]$. Thus $\forall t \ge s \, [\Omega - \Omega_t \le 2^m(\alpha - \alpha_t)]$. So there is a $d$ such that $\forall t \, [\Omega - \Omega_t < d(\alpha - \alpha_t)]$, and hence $\Omega \le_S \alpha$. Contradiction.                       $\dashv$

Finally, we turn to the last part of the theorem. That is, we wish to prove that *if $\alpha$ and $\beta$ are c.e. reals and $\alpha + \beta$ is random then at least one of $\alpha$ and $\beta$ is random.*

This fact will follow easily from a stronger result which shows that, despite the upward density of the Solovay degrees, there is a sense in which the complete Solovay degree is very much above all other Solovay degrees. We begin by noting the following lemma which gives a useful sufficient condition for domination.

LEMMA 31 (Downey, Hirschfeldt, Nies [23]). *Let $f$ be an increasing total computable function and let $k > 0$ be a natural number. Let $\alpha$ and $\beta$ be c.e. reals for which there are infinitely many $s \in \omega$ such that $k(\alpha - \alpha_s) > \beta - \beta_{f(s)}$, but only finitely many $s \in \omega$ such that $k(\alpha_t - \alpha_s) > \beta_{f(t)} - \beta_{f(s)}$ for all $t > s$. Then $\beta \leq_S \alpha$.*

PROOF. By taking $\beta_{f(0)}, \beta_{f(1)}, \ldots$ instead of $\beta_0, \beta_1, \ldots$ as an approximating sequence for $\beta$, we may assume that $f$ is the identity.

By hypothesis, there is an $r \in \omega$ such that for all $s > r$ there is a $t > s$ with $k(\alpha_t - \alpha_s) \leq \beta_t - \beta_s$. Furthermore, there is an $s_0 > r$ such that $k(\alpha - \alpha_{s_0}) > \beta - \beta_{s_0}$. Given $s_i$, let $s_{i+1}$ be the least number greater than $s_i$ such that $k(\alpha_{s_{i+1}} - \alpha_{s_i}) \leq \beta_{s_{i+1}} - \beta_{s_i}$.

Assuming by induction that $k(\alpha - \alpha_{s_i}) > \beta - \beta_{s_i}$, we have

$$k(\alpha - \alpha_{s_{i+1}}) = k(\alpha - \alpha_{s_i}) - k(\alpha_{s_{i+1}} - \alpha_{s_i}) > \beta - \beta_{s_i} - (\beta_{s_{i+1}} - \beta_{s_i})$$
$$= \beta - \beta_{s_{i+1}}.$$

Thus $s_0 < s_1 < \cdots$ is a computable sequence such that $k(\alpha - \alpha_{s_i}) > \beta - \beta_{s_i}$ for all $i \in \omega$.

Now define the computable function $g$ by letting $g(n)$ be the least $s_i$ that is greater than or equal to $n$. Then $\beta - \beta_{g(n)} < k(\alpha - \alpha_{g(n)}) \leq k(\alpha - \alpha_n)$ for all $n \in \omega$, and hence $\beta \leq_S \alpha$.                                    ⊣

We finish the proof of Theorem 28 (iv) by establishing the following.

THEOREM 32 (Downey, Hirschfeldt, Nies [23]). *Let $\alpha$ and $\beta$ be c.e. reals, let $f$ be an increasing total computable function, and let $k > 0$ be a natural number. If $\beta$ is random and there are infinitely many $s \in \omega$ such that $k(\alpha - \alpha_s) > \beta - \beta_{f(s)}$ then $\alpha$ is random.*

PROOF. (sketch) By taking $\beta_{f(0)}, \beta_{f(1)}, \ldots$ instead of $\beta_0, \beta_1, \ldots$ as an approximating sequence for $\beta$, we may assume that $f$ is the identity. If $\alpha$ is rational then we can replace it with a nonrational computable real $\alpha'$ such that $\alpha' - \alpha'_s \geq \alpha - \alpha_s$ for all $s \in \omega$, so we may assume that $\alpha$ is not rational.

We assume that $\alpha$ is nonrandom and there are infinitely many $s \in \omega$ such that $k(\alpha - \alpha_s) > \beta - \beta_s$, and show that $\beta$ is nonrandom. The idea is to take a Solovay test $A = \{I_i : i \in \omega\}$ such that $\alpha \in I_i$ for infinitely many $i \in \omega$ and use it to build a Solovay test $B = \{J_i : i \in \omega\}$ such that $\beta \in J_i$ for infinitely many $i \in \omega$.

Let

$$U = \{s \in \omega : k(\alpha - \alpha_s) > \beta - \beta_s\}.$$

It is not hard to show that $U$ is $\Delta^0_2$, except in the trivial case in which $\beta \equiv_S \alpha$. Thus a first attempt at building $B$ could be to run the following procedure for all $i \in \omega$ in parallel. Look for the least $t$ such that there is an $s < t$ with $s \in U[t]$ and $\alpha_s \in I_i$. If there is more than one number $s$ with this property then choose the least among such numbers. Begin to add the intervals

(1)    $[\beta_s, \beta_s + k(\alpha_{s+1} - \alpha_s)], [\beta_s + k(\alpha_{s+1} - \alpha_s), \beta_s + k(\alpha_{s+2} - \alpha_s)], \ldots$

to $B$, continuing to do so as long as $s$ remains in $U$ and the approximation of $\alpha$ remains in $I_i$. If the approximation of $\alpha$ leaves $I_i$ then end the procedure. If $s$ leaves $U$, say at stage $u$, then repeat the procedure (only considering $t \geq u$, of course).

If $\alpha \in I_i$ then the variable $s$ in the above procedure eventually assumes a value in $U$. For this value, $k(\alpha - \alpha_s) > \beta - \beta_s$, from which it follows that $k(\alpha_u - \alpha_s) > \beta - \beta_s$ for some $u > s$, and hence that $\beta \in [\beta_s, \beta_s + k(\alpha_u - \alpha_s)]$. So $\beta$ must be in one of the intervals (1) added to $B$ by the above procedure.

Since $\alpha$ is in infinitely many of the $I_i$, running the above procedure for all $i \in \omega$ guarantees that $\beta$ is in infinitely many of the intervals in $B$. The problem is that we also need the sum of the lengths of the intervals in $B$ to be finite, and the above procedure gives no control over this sum, since it could easily be the case that we start working with some $s$, see it leave $U$ at some stage $t$ (at which point we have already added to $B$ intervals whose lengths add up to $\alpha_{t-1} - \alpha_s$), and then find that the next $s$ with which we have to work is much smaller than $t$. Since this could happen many times for each $i \in \omega$, we would have no bound on the sum of the lengths of the intervals in $B$.

This problem would be solved if we had an infinite computable subset $T$ of $U$. For each $I_i$, we could look for an $s \in T$ such that $\alpha_s \in I_i$, and then begin to add the intervals (1) to $B$, continuing to do so as long as the approximation of $\alpha$ remained in $I_i$. (Of course, in this easy setting, we could also simply add the single interval $[\beta_s, \beta_s + k \, \mathrm{card}\{I\}]$ to $B$.) It is not hard to check that this would guarantee that if $\alpha \in I_i$ then $\beta$ is in one of the intervals added to $B$, while also ensuring that the sum of the lengths of these intervals is less than or equal to $k \, \mathrm{card}\{I_i\}$. Following this procedure for all $i \in \omega$ would give us the desired Solovay test $B$. Unless $\beta \leq_S \alpha$, however, there is no infinite computable $T \subseteq U$, so we use Lemma 31 to obtain the next best thing. Let

$$S = \{s \in \omega : \forall t > s \, (k \, (\alpha_t - \alpha_s) > \beta_t - \beta_s)\}.$$

If $\beta \leq_S \alpha$ then $\beta$ is nonrandom, so by Lemma 31, we may assume that $S$ is infinite. Furthermore, $S$ is co-c.e. by definition, but it has the additional useful property that if a number $s$ leaves $S$ at stage $t$ then so do all numbers in the interval $(s, t)$.

To construct $B$, we run the following procedure $P_i$ for all $i \in \omega$ in parallel. Note that $B$ is a multiset, so we are allowed to add more than one copy of a given interval to $B$.

1. Look for an $s \in \omega$ such that $\alpha_s \in I_i$.
2. Let $t = s + 1$. If $\alpha_t \notin I_i$ then terminate the procedure.

3. If $s \notin S[t]$ then let $s = t$ and go to step 2. Otherwise, add the interval

$$[\beta_s + k(\alpha_{t-1} - \alpha_s), \beta_s + k(\alpha_t - \alpha_s)]$$

to $B$, increase $t$ by one, and repeat step 3.

This concludes the construction of $B$. It is not hard to show that the sum of the lengths of the intervals in $B$ is finite and that $\beta$ is in infinitely many of the intervals in $B$.                                                                   (sketch) ⊣

So we finally get to prove Theorem 28 (iv) that if $\alpha^0$ and $\alpha^1$ are c.e. reals such that $\alpha^0 + \alpha^1$ is random then at least one of $\alpha^0$ and $\alpha^1$ is random.

Let $\beta = \alpha^0 + \alpha^1$. For each $s \in \omega$, either $3(\alpha^0 - \alpha_s^0) > \beta - \beta_s$ or $3(\alpha^1 - \alpha_s^1) > \beta - \beta_s$, so for some $i < 2$ there are infinitely many $s \in \omega$ such that $3(\alpha^i - \alpha_s^i) > \beta - \beta_s$. By Theorem 32, $\alpha^i$ is random.                    ⊣

We point out that Theorem 28 only applies to c.e. reals. Consider, for instance, if $\Omega = .a_0 a_1 \ldots$ and we put $\alpha = .a_0 0 a_2 0 a_4 0 \ldots$ and $\beta = .0 a_1 0 a_3 0 \ldots$, then clearly neither $\alpha$ nor $\beta$ can be random yet $\alpha + \beta = \Omega$, but they are not c.e.

Before we leave the Solovay degrees of c.e. reals, we note that the structure must be very complicated.

THEOREM 33 (Downey, Hirschfeldt, Laforte [21]). *The Solovay degrees of c.e. reals have an undecidable first-order theory.*

The proof of Theorem 33 uses Nies's method of interpreting effectively dense Boolean algebras, together with a technical construction of a certain class of (strongly) c.e. reals. Calude and Nies [9] have proven that the random reals are all $wtt$-complete. Very little else is known about the Solovay degrees of c.e. reals.

## §11. Other measures of relative randomness.

A reducibility $\leq$ on reals is a *measure of relative randomness* if it satisfies the *Solovay property*:

If $\beta \leq \alpha$ then $\exists c \, (\forall n \, (K(\beta \restriction n) \leq K(\alpha \restriction n) + c))$.

This can also be expressed for $C$ in place of $K$. $S$-reducibility is a measure of relative randomness, but not the only one, and it has some problems:

restricted to c.e. reals,
too fine,
too uniform.

For instance, one can easily construct a real $\alpha$ which is d.c.e. and is not $S$ above any c.e. real.

To see this, imagine you are building a real $\beta$, making sure that it is noncomputable, and trying to defeat all $\varphi_e, c_e$ potential Solovay reductions. We are slowly making $\beta_s > \beta_t$ for $s > t$. Additionally, we are building a computable nonrational real $\alpha = \lim_s \alpha_s$. At some stage $s$, we get that $\varphi_{e,s}(\beta_t) \downarrow$, and

$$c_e(\beta_s - \beta_t) > \alpha_s - \varphi_{e,s}(\beta_t).$$

Then at stage $s + 1$, we simply make $\beta_{s+1}$ sufficiently close to $\beta_t$ to make

$$c_e\left(\beta_{s+1} - \beta_t\right) < \alpha_s - \varphi_{e,s}\left(\beta_t\right).$$

Thus at the very first place it can, Solovay reducibility fails to be useful for classifying relative complexity.

Even on the c.e. reals Solovay reducibility fails badly to encompass relative complexity. In [22], Downey, Hirschfeldt, and Laforte introduced another measure of relative complexity called *sw-reducibility* (strong weak truth table reducibility):

DEFINITION 9. $\beta \leq_{sw} \alpha$ *if there is a functional* $\Gamma$ *such that* $\Gamma^\alpha = \beta$ *and the use of* $\Gamma$ *is bounded by* $x + c$ *for some* $c$.

It is easy to see that by Lemma 25, for any (not necessarily c.e.) reals $\alpha \leq_{sw} \beta$, for all $n$, and $E = C$ or $E = K$,

$$E\left(\alpha \quad n\right) \leq E\left(\beta \quad n\right) + \mathcal{O}(1).$$

The *sw*-degrees have a number of nice aspects, and $\leq_{sw}$ agrees with $\leq_S$ on the *strongly* c.e. reals. Furthermore if $\alpha$ is a c.e. real which is noncomputable, then there are a noncomputable strongly c.e. real $\beta \leq_{sw} \alpha$, and this is *not* true in general, for $\leq_S$. We have the following theorem.

THEOREM 34 (Downey, Hirschfeldt, Laforte [22]). *There is a noncomputable c.e. real* $\alpha$ *such that all strongly c.e. reals dominated by* $\alpha$ *are computable.*

PROOF. Recall that if we have c.e. reals $\beta \leq_S \alpha$ then there is a c.e. real $\gamma$ and a positive $c \in \mathbb{Q}$ such that $\alpha = c\beta + \gamma$.

Now let $\alpha$ be the noncomputable c.e. real $\alpha$ such that if $A$ presents $\alpha$ then $A$ is computable. We claim that, for this $\alpha$, if $\beta \leq_S \alpha$ is strongly c.e. then $\beta$ is computable.

To verify this claim, let $\beta \leq_S \alpha$ be strongly c.e. We know that there is a positive $c \in \mathbb{Q}$ such that $\alpha = c\beta + \gamma$. Let $k \in \omega$ be such that $2^{-k} \leq c$ and let $\delta = \gamma + (c - 2^{-k})\beta$. Then $\delta$ is a c.e. real such that $\alpha = 2^{-k}\beta + \delta$.

It is easy to see that there exist computable sequences of natural numbers $b_0, b_1, \ldots$ and $d_0, d_1, \ldots$ such that $2^{-k}\beta = \sum_{i \in \omega} 2^{-b_i}$ and $\delta = \sum_{i \in \omega} 2^{-d_i}$. Furthermore, since $\beta$ is strongly c.e., so is $2^{-k}\beta$, and hence we can choose $b_0, b_1, \ldots$ to be pairwise distinct, so that the $n$th bit of the binary expansion of $2^{-k}\beta$ is 1 if and only if $n = b_i$ for some $i$.

Since $\sum_{i \in \omega} 2^{-b_i} + \sum_{i \in \omega} 2^{-d_i} = 2^{-k}\beta + \delta = \alpha < 1$, Kraft's inequality tells us that there is a prefix-free c.e. set $A = \{\sigma_0, \sigma_1, \ldots\}$ such that $|\sigma_0| = b_0$, $|\sigma_1| = d_0$, $|\sigma_2| = b_1$, $|\sigma_3| = d_1$, etc. Now $\sum_{\sigma \in A} 2^{-|\sigma|} = \sum_{i \in \omega} 2^{-b_i} + \sum_{i \in \omega} 2^{-d_i} = \alpha$, and thus $A$ presents $\alpha$.

By our choice of $\alpha$, this means that $A$ is computable. But now we can compute the binary expansion of $2^{-k}\beta$ as follows. Given $n$, compute the number $m$ of strings of length $n$ in $A$. If $m = 0$ then $b_i \neq n$ for all $i$, and hence the $n$th bit of binary expansion of $2^{-k}\beta$ is 0. Otherwise, run through the $b_i$ and $d_i$ until either $b_i = n$ for some $i$ or $d_{j_1} = \cdots = d_{j_m} = n$ for some $j_1 < \cdots < j_m$. By the definition of $A$, one of the two cases must happen. In the first case, the $n$th bit of the binary expansion of $2^{-k}\beta$ is 1. In the second case, $b_i \neq n$ for all $i$, and hence the $n$th bit of the binary expansion of $2^{-k}\beta$ is 0. Thus $2^{-k}\beta$ is computable, and hence so is $\beta$.    ⊣

We remark that $sw$-reducibility is also *bad* in many ways. For instance, the $sw$-degrees of c.e. reals do not form a semilattice! (Downey, Hirschfeldt, Laforte [22]). It is unknown if $\Omega$ is $sw$-complete (for c.e. reals), that being the analog of Slaman's theorem. Furthermore $sw$ and $S$ are incomparable.

We would like a measure of relative randomness combining the best of $S$-reducibility and $sw$-reducibility.

Both $S$-reducibility and $sw$-reducibility are uniform in a way that relative initial-segment complexity is not. This makes them too strong, in a sense, and it is natural to wish to investigate nonuniform versions of these reducibilities. Motivated by this consideration, as well as by the problems with $sw$-reducibility, we introduce another measure of relative randomness, called relative $K$-reducibility, which can be seen as a nonuniform version of both $S$-reducibility and $sw$-reducibility, and which combines many of the best features of these reducibilities. Its name derives from a characterization, discussed below, which shows that there is a very natural sense in which it is an *exact* measure of relative randomness.

DEFINITION 10. *Let $\alpha$ and $\beta$ be reals. We say that $\beta$ is relative K reducible (rK-reducible) to $\alpha$, and write $\beta \leq_{rK} \alpha$, if there exist a partial computable binary function $f$ and a constant $k$ such that for each $n$ there is a $j \leq k$ for which $f(\alpha \restriction n, j)\!\downarrow = \beta \restriction n$.*

Clearly $\leq_{rK}$ is transitive. It might seem like a weird definition at first, but the actual motivation came from the consideration of Lemma 25. There we argued that if two strings are very close and of the same length then they have essentially the same complexity no matter whether we use $K$ or $C$. Note that $sw$-reducibility gives a method of taking an initial segment of length $n$ of $\beta$ to one of length $n - c$ of $\alpha$. However, it would be enough to take some string $k$—close to an initial segment or $\beta$ to one similarly close to one of $\alpha$. This idea gives a notion equivalent to $rK$-reducibility and leads to the definition above.

There are, in fact, several characterizations of $rK$-reducibility, each revealing a different facet of the concept. We mention three, beginning with a "relative entropy" characterization whose proof is quite straightforward. For a c.e. real $\beta$ and a fixed computable approximation $\beta_0, \beta_1, \ldots$ of $\beta$, we will let the mind-change function $m(\beta, n, s, t)$ be the cardinality of

$$\{u \in [s, t] \mid \beta_u \restriction n \neq \beta_{u+1} \restriction n\}.$$

LEMMA 35 ([22]). *Let $\alpha$ and $\beta$ be c.e. reals. The following condition holds if and only if $\beta \leq_{rK} \alpha$. There are a constant $k$ and computable approximations*

$\alpha_0, \alpha_1, \ldots$ and $\beta_0, \beta_1, \ldots$ of $\alpha$ and $\beta$, respectively, such that for all $n$ and $t > s$, if $\alpha_t \restriction n = \alpha_s \restriction n$ then $m(\beta, n, s, t) \leq k$.

The following is a more analytic characterization of $rK$-reducibility, which clarifies its nature as a nonuniform version of both $S$-reducibility and $sw$-reducibility.

LEMMA 36 ([22]). *For any reals $\alpha$ and $\beta$, the following condition holds if and only if $\beta \leq_{rK} \alpha$. There is a constant $c$ and a partial computable function $\varphi$ such that for each $n$ there is a $\tau$ of length $n + c$ with $|\alpha - \tau| \leq 2^{-n}$ for which $\varphi(\tau)\downarrow$ and $|\beta - \varphi(\tau)| \leq 2^{-n}$.*

PROOF. First suppose that $\beta \leq_{rK} \alpha$ and let $f$ and $k$ be as in Definition 10. Let $c$ be such that $2^c \geq k$ and define the partial computable function $\varphi$ as follows. Given a string $\sigma$ of length $n$, whenever $f(\sigma, j)\downarrow$ for some new $j \leq k$, choose a new $\tau \supseteq \sigma$ of length $n+c$ and define $\varphi(\tau) = f(\sigma, j)$. Then for each $n$ there is a $\tau \supseteq \alpha \restriction n$ such that $\varphi(\tau)\downarrow = \beta \restriction n$. Since $|\alpha - \tau| \leq |\alpha - \alpha \restriction n| \leq 2^{-n}$ and $|\beta - \beta \restriction n| \leq 2^{-n}$, the condition holds.

Now suppose that the condition holds. For a string $\sigma$ of length $n$, let $S_\sigma$ be the set of all $\mu$ for which there is a $\tau$ of length $n + c$ with $|\sigma - \tau| \leq 2^{-n+1}$ and $|\mu - \varphi(\tau)| \leq 2^{-n+1}$. It is easy to check that there is a $k$ such that $\mathrm{card}\{S_\sigma\} \leq k$ for all $\sigma$. So there is a partial computable binary function $f$ such that for each $\sigma$ and each $\mu \in S_\sigma$ there is a $j \leq k$ with $f(\sigma, j)\downarrow = \mu$. But, since for any real $\gamma$ and any $n$ we have $|\gamma - \gamma \restriction n| \leq 2^{-n}$, it follows that for each $n$ we have $\beta \restriction n \in S_{\alpha \restriction n}$. Thus $f$ and $k$ witness the fact that $\beta \leq_{rK} \alpha$. $\dashv$

The most interesting characterization of $rK$-reducibility (and the reason for its name) is given by the following result which shows that there is a very natural sense in which $rK$-reducibility is an exact measure of relative randomness. Recall that the prefix-free complexity $K(\tau \mid \sigma)$ of $\tau$ *relative to* $\sigma$ is the length of the shortest string $\mu$ such that $M^\sigma(\mu)\downarrow = \tau$, where $M$ is a fixed self-delimiting universal computer. (Similarly for $C$.)

THEOREM 37 ([22]). *Let $\alpha$ and $\beta$ be reals. Then $\beta \leq_{rK} \alpha$ if and only if there is a constant $c$ such that $K(\beta \restriction n \mid \alpha \restriction n) \leq c$ for all $n$. (And $C(\beta \restriction n \mid \alpha \restriction n) \leq c$.)*

PROOF. We give the argument for $K$. First suppose that $\beta \leq_{rK} \alpha$ and let $f$ and $k$ be as in Definition 10. Let $m$ be such that $2^m \geq k$ and let $\tau_0, \ldots, \tau_{2^m-1}$ be the strings of length $m$. Define the prefix-free machine $N$ to act as follows with $\sigma$ as an oracle. For all strings $\mu$ of length not equal to $m$, let $N^\sigma(\mu)\uparrow$. For each $i < 2^m$, if $f(\sigma, i)\downarrow$ then let $N^\sigma(\tau_i)\downarrow = f(\sigma, i)$, and otherwise let $N^\sigma(\tau_i)\uparrow$. Let $e$ be the coding constant of $N$ and let $c = e + m$. Given $n$, there exists a $j \leq k$ for which $f(\alpha \restriction n, j)\downarrow = \beta \restriction n$. For this $j$ we have $N^{\alpha \restriction n}(\tau_j)\downarrow = \beta \restriction n$, which implies that $K(\beta \restriction n \mid \alpha \restriction n) \leq |\tau_j| + e \leq c$.

Now suppose that $K(\beta \restriction n \mid \alpha \restriction n) \leq c$ for all $n$. Let $\tau_0, \ldots, \tau_k$ be a list of all strings of length less than or equal to $c$ and define $f$ as follows. For a string $\sigma$ and a $j \le k$, if $M^\sigma(\tau_j)\!\downarrow$ then $f(\sigma, j)\!\downarrow = M^\sigma(\tau_j)$, and otherwise $f(\sigma, j)\!\uparrow$. Given $n$, since $K(\beta \restriction n \mid \alpha \restriction n) \le c$, it must be the case that $M^{\alpha \restriction n}(\tau_j)\!\downarrow = \beta \restriction n$ for some $j \le k$. For this $j$ we have $f(\alpha \restriction n, j)\!\downarrow = \beta \restriction n$. Thus $\beta \le_{rK} \alpha$.      ⊣

An immediate consequence of this result is that $rK$-reducibility satisfies the Solovay property.

COROLLARY 38. *If $\beta \le_{rK} \alpha$ then there is a constant $c$ such that $K(\beta \restriction n) \le K(\alpha \restriction n) + c$ for all $n$.*

It is not hard to check that the converse of this corollary is not true in general, but the following question is natural.

QUESTION. *Let $\alpha$ and $\beta$ be c.e. reals such that, for some constant $c$, we have $K(\beta \restriction n) \le K(\alpha \restriction n) + c$ for all $n$. Does it follow that $\beta \le_{rK} \alpha$? Does it even follow that $\beta \le_T \alpha$? (See Theorem 41.)*

We will look at this very interesting question in the next section.

THEOREM 39 ([22]). *Let $\alpha$ and $\beta$ be c.e. reals. If $\beta \le_S \alpha$ or $\beta \le_{sw} \alpha$, then $\beta \le_{rK} \alpha$.*

COROLLARY 40. *A c.e. real $\alpha$ is $rK$-complete if and only if it is random.*

Despite the nonuniform nature of its definition, $rK$-reducibility implies Turing reducibility.

THEOREM 41 (Downey, Hirschfeldt, Nies [22]). *If $\beta \le_{rK} \alpha$ then $\beta \le_T \alpha$.*

PROOF. Let $k$ be the least number for which there exists a partial computable binary function $f$ such that for each $n$ there is a $j \le k$ with $f(\alpha \restriction n, j)\!\downarrow = \beta \restriction n$. There must be infinitely many $n$ for which $f(\alpha \restriction n, j)\!\downarrow$ for all $j \le k$, since otherwise we could change finitely much of $f$ to contradict the minimality of $k$. Let $n_0 < n_1 < \cdots$ be an $\alpha$-computable sequence of such $n$. Let $T$ be the $\alpha$-computable subtree of $2^\omega$ obtained by pruning, for each $i$, all the strings of length $n_i$ except for the values of $f(\alpha \restriction n_i, j)$ for $j \le k$.

If $\gamma$ is a path through $T$ then for all $i$ there is a $j \le k$ such that $\gamma$ extends $f(\alpha \restriction n_i, j)$. Thus there are at most $k$ many paths through $T$, and hence each path through $T$ is $\alpha$-computable. But $\beta$ is a path through $T$, so $\beta \le_T \alpha$.      ⊣

Notice that, since any computable real is obviously $rK$-reducible to any other real, the above theorem shows that the computable reals form the least $rK$-degree.

Structurally, the $rK$-degrees of c.e. reals are nicer than the $sw$-degrees of c.e. reals.

THEOREM 42 (Downey, Hirschfeldt, Laforte [22]).

(i) *The $rK$-degrees of c.e. reals form an upper semilattice with least degree that of the computable sets and highest degree that of $\Omega$.*

(ii) *The join of the rK-degrees of the c.e. reals $\alpha$ and $\beta$ is the rK-degree of $\alpha + \beta$.*

(iii) *For any rK-degrees $a < b$ of c.e. reals there is an rK-degree $c$ of c.e. reals such that $a < c < b$.*

(iv) *For any rK-degrees $a < b < \deg_{rK}(\Omega)$ of c.e. reals, there are rK-degrees $c_0$ and $c_1$ of c.e. reals such that $a < c_0, c_1 < b$ and $c_0 \vee c_1 = b$.*

(v) *For any rK-degrees $a, b < \deg_{rK}(\Omega)$ of c.e. reals, $a \vee b < \deg_{rK}(\Omega)$.*

PROOF. We prove only (i); the remainder of the parts are proved by analogous methods to those used for $\leq_S$. All that is left to show is that addition is a join. Since $\alpha, \beta \leq_S \alpha + \beta$, it follows that $\alpha, \beta \leq_{rK} \alpha + \beta$. Let $\gamma$ be a c.e. real such that $\alpha, \beta \leq_{rK} \gamma$. Then Lemma 35 implies that $\alpha + \beta \leq_{rK} \gamma$, since for any $n$ and $s < t$ we have $m(\alpha + \beta, n, s, t) \leq 2(m(\alpha, n, s, t) + m(\beta, n, s, t)) + 1$.  $\dashv$

We remark that the remaining part of Theorem 28 was that $\leq_S$ is distributive on the c.e. reals. This is open at present.

We see that $rK$-reducibility shares many of the nice structural properties of $S$-reducibility on the c.e. reals while still being a reasonable reducibility on non-c.e. reals. Together with its various characterizations, especially the one in terms of relative $K$-complexity of initial segments, this makes $rK$-reducibility a tool with great potential in the study of the relative randomness of reals. As one would expect, little else is known about the structure of $rK$ degrees.

We remark that the methods of this section have been used by Downey, Hirschfeldt, and Laforte [22] to prove that the $K$-degrees of c.e. reals are dense.

§12. $\leq_K$, $\leq_C$, and $\leq_T$. Let us return to the question below we deferred from the last section.

QUESTION. *Let $\alpha$ and $\beta$ be c.e. reals such that, for some constant $c$, we have $K(\beta \restriction n) \leq K(\alpha \restriction n) + c$ for all $n$. Does it follow that $\beta \leq_{rK} \alpha$? Does it even follow that $\beta \leq_T \alpha$?*

Although it might seem at first that the answer to this question should obviously be negative, Theorem 43 would seem to indicate that any counterexample would probably have to be quite complicated and gives us hope for a positive answer.

THEOREM 43 (Downey, Hirschfeldt, Laforte [22]). *Let $\alpha$ and $\beta$ be c.e. reals such that $\liminf_n K(\alpha \restriction n) - K(\beta \restriction n) = \infty$. Then $\beta <_{sw} \alpha$.*

PROOF. Let $c_\alpha(n)$ be the least $s$ such that $\alpha_s \restriction n = \alpha \restriction n$, and define $c_\beta(n)$ analogously. Let $M$ be a universal self-delimiting computer and define the self-delimiting computer $N$ as follows. For each $n$, $s$, and $\sigma$, if $M(\sigma)[s]\downarrow = \beta_s \restriction n$ and $N(\sigma)$ has not been defined before stage $s$ then let $N(\sigma)\downarrow = \alpha_s \restriction n$. Let $e$ be the coding constant of $N$. For each $n$, if $c_\beta(n) \geq c_\alpha(n)$ then $\forall \sigma(M(\sigma)\downarrow = \beta \restriction n \Rightarrow N(\sigma)\downarrow = \alpha \restriction n)$, which implies that $K(\alpha \restriction n) \leq K(\beta \restriction n) + e$.

Thus our hypothesis implies that $c_\beta(n) < c_\alpha(n)$ for almost all $n$, which clearly implies that $\beta \leq_{sw} \alpha$. We note that $\alpha \not\leq_{sw} \beta$, so $\beta <_{sw} \alpha$.                    ⊣

Stephan [62] has shown that the theorem above has limited use because it is hard to satisfy the hypotheses of the theorem. Let $c_\alpha$ denote the computation function of $\alpha$: $c_\alpha(x)$ is the least $s \geq x$ with $\alpha \upharpoonright x = \alpha_s \upharpoonright x$. The following lemma is easy.

LEMMA 44. *Let $A$ be a c.e. set (or c.e. real). Suppose that $c_A$ dominates all partial computable functions. Then $A$ is wtt-complete.*

THEOREM 45 (Stephan [62]). *Suppose that $\alpha$ and $\beta$ satisfy the hypotheses of Theorem 43. Then $\alpha$ is wtt-complete.*

PROOF. Given $M$ a prefix-free universal machine and $\varphi$ a partial computable function, define

$$\widehat{M}(a\tau) = \begin{cases} M(\tau) & \text{if } a = 0, \\ \alpha_{\varphi(|M(\tau)|)} \upharpoonright M(|\tau|) & \text{if } a = 1, M(\tau)\downarrow, \text{ and } \varphi(|M(\tau)|)\downarrow, \\ \uparrow & \text{otherwise.} \end{cases}$$

Then $\widehat{M}$ is also a prefix-free universal machine. If $c_\alpha$ does not dominate $\varphi$, then there are infinitely many $n$ with $\varphi(n)\downarrow > c_\alpha(n)$, and thus

$$K_{\widehat{M}}(\alpha \upharpoonright n) = \min \left\{ K_{\widehat{M}}(\sigma) : |\sigma| = n \right\}.$$

That is, $\alpha \upharpoonright n$ has minimum $K$-complexity of all strings of length $|\alpha \upharpoonright n| = n$. Consequently,

$$K_{\widehat{M}}(\alpha \upharpoonright n) \leq K_{\widehat{M}}(\beta \upharpoonright n),$$

for these $n$, and $\liminf(K_{\widehat{M}}(\alpha \upharpoonright n) - K_{\widehat{M}}(\beta \upharpoonright n)) \leq 0$.

So suppose that $\alpha$ is not *wtt*-complete. Then there is some partial computable function not dominated by $c_\alpha$. So there is no $\beta$ with

$$\liminf \left( K_{\widehat{M}}(\alpha \upharpoonright n) - K_{\widehat{M}}(\beta \upharpoonright n) \right) = \infty.$$                    ⊣

Stephan has clarified the situation for the relationship between $\leq_K$ and $\leq_T$.

THEOREM 46 (Stephan [62]). *Suppose that we have c.e. $\alpha, \beta$ with $\alpha \leq_C \beta$. Then $\alpha \leq_T \beta$.*

PROOF. So we suppose that there is a $c$ such that for all $n$,

$$K(\alpha \upharpoonright n) \leq K(\beta \upharpoonright n) + c.$$

If $\beta \equiv_T \emptyset'$, then there is nothing to prove. So we suppose that $\beta <_T \emptyset'$. In that case, for any total $\beta$-computable function $g$, we know

$$\exists^\infty x \left[ x \in K - K_{g(x)} \right].$$

Let $\psi(x)$ be the partial computable function of the least stage that $x \in K_s$. Then there are infinitely many $x$ with $\psi(x)\!\downarrow\, > g(x)$. So let $g$ be the computation function of $\beta$. Then there is an infinite $D \leq_\beta$ with $D \subset K$ and $\psi(x) > g(x)$ for all $x \in D$.

For any $x \in D$ we have the following program:

$$\varphi_e(x) = \beta_{\psi(x)} \restriction x.$$

For this set of $x$ we have $C(\beta \restriction x \mid x) \leq e$, and hence

$$K(\alpha \restriction x \mid x) \leq e + \mathcal{O}(1).$$

Now relativizing Loveland's theorem below, we see that $\alpha \leq_T \beta$. ⊣

The missing ingredient is an old result of Loveland (actually this is stated in slightly generalized form).

THEOREM 47 (Loveland [49]). *Suppose that there is $e$ and an infinite computable set $A$ such that for all $x \in A$, $C(\alpha \restriction x \mid x) \leq e$. Then $\alpha$ is computable. Moreover for each $e$ there are only finitely many $\alpha$ with $C(\alpha \restriction x \mid x) \leq e$ for all $x \in A$.*

PROOF. Assume for all $x \in A$, $C(\alpha \restriction x \mid x) \leq e$. Note that there are at most $f = \mathcal{O}(2^e)$ many programs of size $e$ or less. There will be a maximum number $g \leq f$ which will be invoked infinitely often. That is, there is a maximum $g$, and programs $p_1, \ldots, p_g$, such that for infinitely many $n \in A$, $U(p_i, n) = \sigma$ with $|\sigma| = n$. Working above some finite initial segment of $\alpha$ we may assume that there are never more than $g$ many such programs for any $n \in A$. We can compute an infinite increasing collection of $n_i \in A$ such that $g$ is achieved on $n_i$. We can use the $n_i$ to generate a $\Pi^0_1$ class from a computable tree $T$,

$$\tilde{T} = \Big\{ \sigma \colon |\sigma| = n_i \wedge \forall v \preceq \sigma \exists j$$
$$\leq i\big(|v| = n_j \to \exists p_j(|p_j| \leq e \wedge U(p_j, n_j) = v)\big) \Big\}.$$

$T$ is the downward closure of $\tilde{T}$. Then $T$ is clearly computable. The paths through $T$ form a $\Pi^0_1$ class and correspond to the reals $\alpha$ with $C(\alpha \restriction n, n) \leq e$ for all $n \in A$. Noticing that $T$ has at most $g$ many paths, we see that all such $\alpha$ are computable. ⊣

One of the keys to the above is the uniformity implicit in $C(x \restriction n \mid y \restriction n)$. A much more interesting theorem is the following of Chaitin which indicates a hidden uniformity in $C$.

THEOREM 48 (Chaitin [10]). *Suppose that $C(\alpha \restriction n) \leq C(n) + \mathcal{O}(1)$ for all $n$ (for an infinite computable set of $n$), or $C(\alpha \restriction n) \leq \log n + \mathcal{O}(1)$, for all $n$. Then $\alpha$ is computable (and conversely). Furthermore for a given constant $\mathcal{O}(1) = d$, there are only finitely many $(\mathcal{O}(2^d))$ such $x$.*

The proof of Chaitin's theorem involves a lemma of independent interest. The proof below is along the lines of Chaitin's, but we hope that it is somewhat less challenging than the original.

Let $D: \sum^* \mapsto \sum^*$ be partial computable. Then a *D-description* of $\sigma$ is a pre-image of $\sigma$.

LEMMA 49 (Chaitin [10]). *Let* $f(d) = 2^{(d+c)}$, $c = c_{d,D}$ *to be determined. Then for each* $\sigma \in \sum^*$,

$$|\{q: D(q) = \sigma \wedge |q| \leq C(\sigma) + d\}| \leq f_D(d).$$

*That is, the number of D-descriptions of length* $\leq C(\sigma) + d$ *is bounded by an absolute constant depending upon* $d, D$ *alone (and not upon* $\sigma$*).*

Note that this applies in the special case that $D$ is the universal machine.

PROOF. Let $\sigma$ be given, and $k = C(\sigma) + d$. For each $m$ there are at most $2^{k-m} - 1$ strings with $\geq 2^m$ $D$-descriptions of length $\leq k$, since there are $2^k - 1$ strings in total. Given $k, m$ we can effectively list strings $\sigma$ with $\geq 2^m$ $D$-descriptions of length $\leq k$, uniformly in $k, m$. (Wait until you see $2^m$ $q$s of length $\leq k$ with $D(q) = v$ and then put $v$ on the list $L_{k,m}$.) The list $L_{k,m}$ has length $\leq 2^{k-m}$.

If $\sigma$ has $\geq 2^m$ $D$-descriptions of length $\leq k$, then it is given by

$m$

a string $q$ of length $2^{k-m}$,

the latter indicating the position of $\sigma$ in $L_{k,m}$. This description has length bounded by $\log m + k - m + c$ where $c$ depends only upon $D$. If we choose $m$ large enough so that $\log m + k - m + c < k - d$, we can then get a description of $\sigma$ of length $< k - d = C(\sigma)$. If we let $f(d)$ be $2^n$ where $n$ is the least $m$ with $\log m + c + d < m$ then we are done.                                ⊣

The next lemma tells us that there are relatively few string with short descriptions and the number depends on $d$ alone.

LEMMA 50 (Chaitin [10]). *There is a computable* $h$ *depending only on* $d$ $(h(d) = \mathcal{O}(2^d))$ *such that, for all* $n$,

$$|\{\sigma: C(\sigma) \leq C(n) + d\}| \leq h(d).$$

PROOF. Consider the partial computable function $D$ defined via $D(p)$ is the unary representation of $U(p)$. Then let $h(d) = f_D(d)$, with $f$ given by the previous lemma. Suppose that $C(\sigma) \leq C(n) + d$, and pick the shortest $p$ with $U(p) = \sigma$. Then $p$ is a $D$-description of $n$ and $|p| \leq C(n) + d$. Thus there are at most $f(d)$ many $p$s, and hence $\sigma$s.                                ⊣

PROOF OF THEOREM 48 CONCLUDED. Let

$$T = \{\sigma: \forall p \subseteq \sigma (C(p) \leq \log|p| + d)\}.$$

If $n$ is random then $C(n) = \log n + c$, so that by Lemma 49 above, the number of strings in $T$ of length $n$ is $\leq h(d)$. Taking the maximum number $\leq h(d)$

attained infinitely often, we can then construct a computable subtree of the c.e. tree $T$, upon which $x$ must be a path. Note that the number of paths is bounded by $h(d)$. ⊣

It is still an open question whether, for c.e. reals, $\leq_C$ implies $\leq_{rC}$, although the answer is "surely not".

The situation for $\leq_K$ is quite different. The argument of Stephan above shows that $\alpha \leq_K \beta$ implies that for all $x \in D$, $K(\beta \upharpoonright x \mid x) \leq e$, and hence $K(\beta \upharpoonright x) \leq e + K(x) + \mathcal{O}(1)$, for this set of $x$. All would be sweet if the following statement, true for $C$, was also true for $K$: $K(\alpha \upharpoonright x) \leq K(x) + \mathcal{O}(1)$ for all (a computable set of) $x$, implied that $\alpha$ is computable. Chaitin observed using a relativized form of Loveland's observation that

$$H(\alpha \upharpoonright x) \leq K(x) + \mathcal{O}(1) \text{ implies } \alpha \leq_T \emptyset'.$$

*Surprisingly we cannot replace $\emptyset'$ by $\emptyset$ for $K$. That is, even though $\alpha$ looks identical to $\omega$ we cannot conclude that $\alpha$ is computable even for strongly c.e. reals $\alpha$.*

This was proved by Solovay in his 1974 manuscript. The proof there is very complicated and only constructs a $\Delta_2^0$ real. For the remainder of the section we will prove a slight generalization of Solovay's theorem. In fact we will give *two* proofs as the result is so interesting and each proof yields a different technique. In particular, the second is a modification of Solovay's original proof which contains an important lemma of independent interest. The first is short and easy once you have found it. Both are due to Downey, Hirschfeldt, and Nies.

THEOREM 51 (Downey, Hirschfeldt, Nies, after Solovay [61]). *There is a c.e. noncomputable set $A$ such that, for all $n$,*

$$K(A \upharpoonright n) \leq K(n) + \mathcal{O}(1).$$

PROOF. Whereas the proof below is easy, it is slightly hard to see why it works. So, by way of motivation, suppose that we were asked to "prove" that the set $B = \{0^n : n \in \omega\}$ had the same complexity as $\omega = \{1^n : n \in \omega\}$. A complicated way to do this would be for us to build our own prefix-free machine $M$ whose only job was to compute initial segments of $B$. The idea was if the universal machine $U$ enumerated $\langle \sigma, 1^n \rangle$, then in our machine we would enumerate $\langle \sigma, 0^n \rangle$. Notice that, in fact, using Kraft-Chaitin it would be enough to build $M$ *implicitly* enumerating the length axiom (or "requirement") $\langle |\sigma|, n \rangle$. We are guaranteed that

$$\sum_{\sigma \in \text{dom}(U)} 2^{-|\sigma|} = \sum_{\tau \in \text{dom}(M)} 2^{-|\tau|} < 1.$$

Hence Kraft-Chaitin applies.

Note also that we could, for convenience, as we do in the main construction, use a string of length $|\sigma| + 1$, in which case we would force

$$\sum_{\tau \in \text{dom}(M)} 2^{-|\tau|} < \frac{1}{2}.$$

The idea is the following. We will build a noncomputable c.e. set $A$ in place of $B$ and, as above, we will slavishly follow $U$ on $n$ in the sense that whenever $U$ enumerates, at stage $s$, a shorter $\sigma$ with $U(\sigma) = n$, then we will, in our machine $M$, enumerate $\langle \tau, A_s \restriction n \rangle$, where $|\tau| = |\sigma| + 1$. To make $A$ noncomputable, we will also sometimes make $A_s \restriction n \neq A_{s+1} \restriction n$. Then for each $j$ with $n \leq j \leq s$, we will for the currently shortest string $\sigma_j$ computing $j$, also need to put into $M$,

$$\langle \tau_j, A_{s+1} \restriction j \rangle.$$

This construction works by making this quantity small. We are ready to define $A$:

$$A = \left\{ \langle e, n \rangle : W_{e,s} \cap A_s = \emptyset \wedge \langle e, n \rangle \in W_{e,s} \wedge \sum_{\langle e,n \rangle \leq j \leq s} 2^{-K(j)[s]} < 2^{-(e+1)} \right\}.$$

Then clearly $A$ is c.e. It is noncomputable since the $K$-complexity of $P_n = \{m : m \geq n\}$ tends to zero as $n \to \infty$, and finally, $M$ is a Chaitin machine, since the errors are bounded by $\sum_e 2^{-(e+1)}$ (once for each $e$), whence

$$\sum_{\sigma \in \text{dom}(M)} 2^{-|\sigma|} < \sum_{\sigma \in \text{dom}(U)} 2^{|\sigma|+1} + \sum_e 2^{-(e+1)} < \frac{1}{2} + \frac{1}{2} = 1.$$

And by force we have for all $n$, $K(A \restriction n) \leq K(n) + \mathcal{O}(1)$.                    ⊣

The second proof is a modification of Solovay's. It is more complicated. The basic idea is to try to let $U$ do the work for us. Specifically, consider the task of constructing a noncomputable $\alpha$ with $K(\alpha \restriction n) \leq K(n) + \mathcal{O}(1)$ for all $n$. The natural idea would be that we would pick some diagonalization place $i$ and wait for $i$, keep it out of $\alpha_s = .A_s$ until $i$ appears in $W_{e,s}$ for the requirement $R_e$ saying $\overline{A} \neq W_e$. Then one would put $i$ into $A_{s+1}$.

Now this stage $s$ could be very late. Furthermore we need that $K(\alpha \restriction i) \leq K(i) + \mathcal{O}(1)$. In the one above we allow ourselves to do this provided that the amount of damage we do is small. Solovay's idea is to only let ourselves do this provided we can keep $K(\alpha \restriction i) \leq K(i)[s]$. The idea is that we will let the opponent do the work for us. Occasionally the opponent must drop the stage $s$ complexity of $i$ to something lower. Our idea is that then at that very stage we can change $A_s(i)$ and the amount of entropy we need will simply follow the universal computer on $i$. Thus the argument looks very easy.

There is a big problem which necessitates all the following. Suppose that $U$ changes its current complexity on $i$ at the stage $t$. At stage $t$, *we have enumerated A up to length t.* Now if we change $A$ on $i$, then for all $t \geq n \geq i$, we *also change A ↾ n.* Hence, for each of these $A ↾ n$ we *also* would need to enumerate some string $\sigma$ computing $A ↾ n$ of the same length as the one computing $n$, and perhaps only $i$ has a new string. Why should $n$?

The key idea of Solovay is that if one looks at appropriately sparse stages of the construction, then one can prove that there will, infinitely often, be stages where *all* of the $n \geq i$ get new shorter descriptions *together*. This is by no means obvious. We turn to the formal proof.

Fix a universal prefix-free machine $U$ and define $K$ relative to $U$. We may assume that $K(n) \leq n + 1$ for all $n$, and hence we adopt the convention that if there is no $\tau$ such that $|\tau| \leq n$ and $U(\tau)[s]\downarrow = n$ then $K(n)[s] = n + 1$.

Let $A$ be any function with primitive recursive graph that dominates all primitive recursive functions (e.g., Ackermann's function) and define $t_0 = 0$ and $t_{n+1} = A(t_n)$. Let $\sigma(i)$ be the largest $j \leq i$ such that $K(n)[t_i] = K(n)[t_{i+1}]$.

THEOREM 52 (Solovay). *For any total computable function $g$ there are infinitely many $i$ such that $g(\sigma(i)) < i$.*

PROOF. Fix a total computable function $g$. Let $G(n)$ be the least $m$ such that $g(m) \geq n$. To show that there are infinitely many $i$ such that $g(\sigma(i)) < i$, it is enough to show that there are infinitely many $i$ such that $G(i) > \sigma(i)$. By increasing $g$ if necessary, we may assume that $g$ has primitive recursive graph and that $g(n) \geq n$ for all $n$, which implies that $G$ is primitive recursive.

For each $k$, let $n_k$ be the least $n$ such that

$$\sum_{G(n) \leq j \leq n} 2^{-K(j)[t_n]} \leq 2^{-2k},$$

which exists since for any $n$ we have

$$\sum_{G(n) \leq j \leq n} 2^{-K(j)[t_n]} < \sum_{j \geq G(n)} 2^{-K(j)},$$

and this last sum goes to 0 as $n$ increases. Note that the graph of the function $k \mapsto t_{n_k}$ is primitive recursive.

Consider the following enumeration of requirements. For each $k > 0$ and each $j \in [G(n_k), n_k]$, enumerate the requirement $\langle j, K(j)[t_{n_k}] - k \rangle$. Since

$$\sum_{k>0} \sum_{G(n_k) \leq j \leq n_k} 2^{-(K(j)[t_{n_k}]-k)} = \sum_{k>0} 2^k \sum_{G(n_k) \leq j \leq n_k} 2^{-(K(j)[t_{n_k}])} \leq \sum_{k>0} 2^k 2^{-2k} = 1,$$

the Kraft-Chaitin Theorem implies that there is a prefix-free $M$ satisfying these requirements. Furthermore, $M$ can be built to satisfy each requirement $\langle j, K(j)[t_{n_k}] - k \rangle$ in time primitive recursive in $t_{n_k}$, and hence before time $t_{n_k+1}$ for sufficiently large $k$.

Thus, for all sufficiently large $k$, we have $H_M(j)[t_{n_k+1}] = K(j)[t_{n_k}] - k$ for all $j \in [G(n_k), n_k]$. This means that, for all sufficiently large $k$, we have $K(j)[t_{n_k+1}] < K(j)[t_{n_k}]$ for all $j \in [G(n_k), n_k]$, which implies that $G(n_k) > \sigma(n_k)$. So there are infinitely many $n$ such that $G(n) > \sigma(n)$.    ⊣

We call a real $\alpha$ with $K(\alpha \upharpoonright n) \leq K(n) + \mathcal{O}(1)$ for all $n$ $K$-trivial. Now we can define two reals,

$$\Lambda = \sum_{i \in \omega} 2^{-2(\sigma(i))},$$

$$\Lambda_C = \sum \left\{ 2^{-(2(\sigma(i))^2 + j)} : i \in \omega \wedge \sigma(i)[t_{i+1}] \leq i \text{ at least } j \text{ times} \right\}.$$

Then $\Lambda$ and $\Lambda_C$ are $K$-trivial, and additionally, the latter is a strongly c.e. real. (For instance, if we assume that $\Lambda$ is computable, then the stage when the first $i$ bits of $\Lambda$ stop moving allows us to define a computable function $g$, after which $\sigma(i)$ cannot drop below $i$, contrary to Theorem 52. Also $\sigma(i)$ can drop below $i$ only $i$ (in fact $\log i$) times, and hence $\Lambda$ exists.)

We remark that the first construction clearly combines with, for instance, permitting: below any c.e. nonzero degree there is a noncomputable c.e. $K$-trivial set. Also, one can use it to construct a promptly simple one, or a variant to avoid a given low c.e. set, using the Robinson trick. However, we do not know if there is a complete $K$-trivial set. An answer either way would be very interesting. In the positive way it says that the relationship between $\leq_K$ and $\leq_T$ fails as badly as it can. In the negative way, then the first proof of Theorem 51 above provides a priority-free (or more precisely "injury-free") solution to Post's problem.[8] While these are known, the construction we give is particularly simple. The Kučera-Terwijn result of the next section is another example of this phenomenom.

§13. Other areas. Naturally in a short course such as this I cannot hope to cover all areas falling under the umbrella of the intrinsic relationship between computability and randomness. In this last section, I will point toward other material of which I am aware, and direct the reader toward the literature. I certainly do not claim completeness here. In particular, I will not even discuss the rich area of resource bounded randomness. (See e.g., Ambos-Spies et al. [2].) Otherwise the reader has Ambos-Spies and Kučera [1], van Lambalgen [65], and Li-Vitanyi [48] as general references.

13.1. $\Pi^0_1$ classes. One natural direction is to examine not necessarily c.e. reals, and perhaps the collection of all random reals. A counting argument shows that the set of random reals has measure 1. Kurtz [46] and Kautz [37] have a lot of material here. A nice observation more or less due to Martin-Löf [51] is that the random reals form a $\Sigma^0_2$ class: an effective union of $\Pi^0_1$ classes.

---

[8]Downey, Hirschfeldt, Nies, and Stephan [24] have recently proven that an $K$-trivial set is never $T$-complete, and hence the above is a priority-free solution to Post's Problem.

To see this let

$$C_k = \{x \in [0, 1] \colon \forall n K(x \restriction n) \geq n - k\}.$$

Then evidently the $C_k$ form $\Pi_1^0$ classes and $x$ is random iff there is a $k$ with $x \in C_k$.

$$RAND = \cup_k C_k.$$

As a consequence, all the apparatus of $\Pi_1^0$ classes apply. There are, for instance, random reals of low degree by the low basis theorem. Kučera [42, 41] has a lot of material here, additionally relating these notions to genericity and other notions such as $DNR$ functions.

**13.2. Martin-Löf lowness.** One very interesting area comes from relativizing the notion of a test. We can define the notion of a Martin-Löf test relative to an oracle, and hence get the class $RAND^X$. Van Lambalgen and Zambella asked if there is a Martin-Löf low set $X$: a set $X$ such that $RAND^X = RAND$. This question has an affirmative answer.

THEOREM 53 (Kučera and Terwijn [44]). *There is a c.e. set A that is Martin-Löf low.*

PROOF. We give an alternative proof to that in [44]. It is clear that there is a primitive recursive function $f$, so that $U_{f(n)}^A$ is the universal Martin-Löf test relative to $A$. Let $I_n^A$ denote the corresponding Solovay test. Then $X$ is $A$-random iff $X$ is in at most finitely many $I_n^A$. We show how to build a $\{J_n \colon n \in \omega\}$, a Solovay test, so that if $(p, q) \in I_n^A$, it is also in $J_n$. This is done by simple copying: if $(p, q) < s$ is in $\cup_{j \leq s} I_j^{A_s}$ is not in $J_i \colon i \in s$, add it. Clearly this "test" has the desired property of covering $I_n^A$. We need to make $A$ so that the "mistakes" are not too big.

The crucial concept comes from Kučera and Terwijn: Let $M_s(y)$ denote the collection of intervals $\{I_n^{A_s} \colon n \leq s\}$ which have $A_s(y) = 0$ in their use function. Then we put $y > 2e$ into $A_{s+1} - A_s$ provided that $e$ is least with $A_s \cap W_{e,s} = \emptyset$, and

$$\mu\left(M_s(y)\right) < 2^{-e}.$$

It is easy to see that this can happen at most once for $e$ and hence the measure of the total mistakes is bounded by $\sum 2^{-n}$ and hence the resulting test is a Solovay test. The only thing we need to prove is that $A$ is noncomputable. This follows since, with priority $e$, whenever we see some $y$ with $\mu(M_s(y)) \geq 2^{-e}$, such $y$ will *not* be added and hence this amount of the $A$-Solovay test will be protected. But since the total measure is bounded by 1, this cannot happen forever. ⊣

There is no known characterization of the degrees of such sets. Clearly the above argument permits and hence each nonzero c.e. degree has a nonzero

Martin-Löf low predecessor. Kučera and Terwijn show that each Martin-Löf low set is in the class $GL_1$: the sets $A$ with $A' \equiv_T A \oplus \emptyset'$.

Intriguingly, there is a *complete* characterization for the Schnorr low sets of the next section.

**13.3. Schnorr lowness.** As noted in the earlier sections, the notion of Martin-Löf randomness is by no means the *only* notion of algorithmic randomness. The choice of it as the "correct" notion is as much philosophical as mathematical. There are a number of competing notions. One that has much support is due to Schnorr.

DEFINITION 11. *A real $x$ is called* Schnorr random *iff it passes all Schnorr tests. A Schnorr test is a Martin-Löf test* $\{U_n : n \in \omega\}$ *such that for all $n$*

$$\mu(U_n) = 2^{-n}.$$

Recall that the idea of a Martin-Löf test was to avoid all sets which were *effectively null*. The difference between a Schnorr and a Martin-Löf test is the relevant level of effectiveness demanded. There are no universal Schnorr tests. Indeed, one can construct a computable real passing a given Schnorr test. Clearly every Martin-Löf random set is Schnorr random. The converse fails.

THEOREM 54 (Schnorr). *There are c.e. reals that are Schnorr random but not Martin-Löf random.*

The idea of the proof is to build out real $\alpha$ and a Martin-Löf test $\{U_i : i \in \omega\}$, by enumerating all partial Schnorr tests, and waiting till the opponent enumerates a lot of his test, then building in the complement. (Thus we can choose to pretend that a partial Schnorr test is not really one until he enumerates to within of the claimed measure of $V_1, V_2, \ldots, V_n$ for some fixed $n$.)

Actually this theorem follows from some recent work of Downey and Griffiths [16]. Downey and Griffiths have shown that every Schnorr random c.e. real is of high c.e. degree. However, they also use a relatively difficult $0''$ argument to prove that there exist incomplete Schnorr random c.e. reals. Since all Martin-Löf random c.e. reals are $T$-complete, such an incomplete Schnorr random c.e. real cannot be Martin-Löf random. We refer the reader to Downey-Griffiths [16] for this and other results here.

Little is known about the degrees of Schnorr random reals. There is no known combinatorial characterization like Chaitin randomness for Schnorr randomness. It seems to the author there is a whole constellation of questions about this and other notions of randomness such as weak randomness of Kurtz [46] awaiting the development of the appropriate technology.

One really nice aspect of Schnorr randomness is that there is a complete characterization of Schnorr low sets. As usual, let $D_x$ denote the $x$th canonical finite set.

THEOREM 55 (Terwijn and Zambella [63]). *A set $X$ is Schnorr low iff there is a computable function $p$ such that, for all functions $g \leq_T X$, there is a function $h$ where, for all $n$,*

(i) $|H_{h(n)}| < p(n)$,

(ii) $g(n) \in D_{h(n)}$.

The proof is nontrivial. It relies in one direction on ideas of Rasonnier [54] on rapid filters for the "mathematical" proof of Shelah's theorem that you cannot take the inaccessible cardinal away from Solovay's construction of a model where every set of reals is Lebesgue measurable. Note that all such Schnorr low degrees are hyperimmune free and hence are *not* below $\mathbf{0}'$. This is quite different from the situation for Martin-Löf lowness as there the set constructed was c.e. It is unknown how the two notions relate. It is unknown if there is a similar characterization for Martin-Löf lowness although one direction works. It is known that not all hyperimmune free degrees are Schnorr low.

**13.4. Computably enumerable sets.** An important subtopic is the complexity of strongly c.e. reals, that is, c.e. sets. We mention only one result here but there are many more, and many open questions. As we have seen, the Kolmogorov complexity of a c.e. set $A \restriction n$ is bounded by $2 \log n - \mathcal{O}(1)$. Solovay asked if this bound was attainable. Certainly for $C$-complexity no c.e. set has initial segment complexity on length $n$ always greater than $2 \log n - \mathcal{O}(1)$.

In a very interesting paper, Kummer [45] proved that there are c.e. *complex* sets.

THEOREM 56 (Kummer [45]). *There is a set $A$ such that there is a constant $c$, with*

$$K(A \restriction n) \geq 2 \log n - c$$

*for infinitely many $n$.*

PROOF. Kummer's proof runs as follows. First we have intervals defined via $t_0 = 0$, $t_{i+1} = 2^{t_i}$ and then $I_i = (t_i, t_{i+1}]$. The Kummer defined

$$f(k) = \sum_{i=t_k+1}^{t_{k+1}} (i - t_k + 1),$$

$$g(k) = \max \left\{ d : 2^d - 1 < f(k) \right\}.$$

Note that $f(k)$ asymptotically approaches $1/2 t_{k+1}^2$ and $g(k)$ approaches $2 \log t_{k+1} - 2$. Then at stage $s + 1$ our action is the following for $k = 0, \ldots s$, if $C(A_s \restriction n) \leq g(k)$ for all $n \in I_k$, put the minimum element in $\overline{A}_s \cap I_k$ into $A_{s+1}$.

Now suppose that $C(A \restriction n) \leq g(k)$ for all $n \in I_k$. Then all of $I_k$ is put into $A$. For a fixed $n$ there are at least $n - t_k + 1$ many strings $\sigma = A \restriction n$

with $|\sigma| = n + 1$ and $C(\sigma) \leq g(k)$. Therefore there are at least $f(k)$ many strings of $C$ complexity at most $g(k)$ and this contradicts the fact that $f(k) > 2^{g(k)+1} - 1$.                                                                            ⊣

Actually Kummer classified the degrees containing complex c.e. sets. From Downey, Jockusch, and Stob [26], a degree **a** is called *array noncomputable* iff for all $g \leq_{tt} \emptyset'$ there is a function $h \leq_T$ **a** not dominated by $g$. The anc degrees form an upward closed class of the c.e. degrees including some low degrees, but such that each c.e. degree has a nonzero array computable predecessor. The array noncomputable degrees capture a notion of "multiple permitting" common to a number of degree constructions. For example, there are the degrees $A \oplus B$ such that $A$ and $B$ are c.e. and have no complete separating set. Ishmukhametov [35] has the following characterization of the (c.e.) array computable degrees.

**a** *is array computable iff there is a computable function $p$ such that, for all $g \leq_T$* **a**, *there is a computable function $h$ such that*

(i) $|W_{h(n)}| < p(n)$, *and*
(ii) $g(n) \in W_{h(n)}$.

The reader might like to compare this characterization with the one for Schnorr low sets. Surely there must be some connection here! I remark that Ishmukhametov [35] used this characterization to prove that the c.e. degrees with strong minimal covers are exactly the array computable ones. Of interest to us here is the following.

THEOREM 57 (Kummer [45]). *A c.e. degree contains a complex set iff it is array noncomputable. Furthermore if the degree is array computable and $A$ is any c.e. set of the degree, then for any $> 1$,*

$$K(A \upharpoonright n) \leq (1 + ) \log n + \mathcal{O}(1).$$

Clearly there are other connections between degree, domination, and complexity.

**Note added, 11 June, 2004.** This paper was written in 2001. In the intervening years there has been remarkable progress in the area. Some of the questions asked in the present paper have been answered. The "trivial" reals have been extensively studied by Nies who showed that they form a "natural" $\Sigma_3^0$ ideal in the Turing degrees, consisting of only low degrees bounded by computably enumerable degrees.

There is insufficent room here to mention all of this exciting progress. Instead, we refer the reader to the following two survey papers, [14] and [25], which report on this and the upcoming book of the author and Denis Hirschfeldt [20].

The original version of this paper used the $H, K$ notation, but, in an area where there seems no standard notation, I have now adopted Li and Vitanyi's notation of $C$ for plain Kolmogorov complexity and $K$ for the prefix-free variety.

## REFERENCES

[1] K. AMBOS-SPIES and A. KUČERA, *Randomness in computability theory*, **Computability theory and its applications** (Cholak, Lempp, Lerman, and Shore, editors), Contemporary Mathematics, vol. 257, 2000, pp. 1–14.

[2] K. AMBOS-SPIES and E. MAYORDOMO, *Resource bounded measure and randomness*, **Complexity, logic and recursion theory** (A. Sorbi, editor), Marcel-Decker, New York, 1997, pp. 1–48.

[3] K. AMBOS-SPIES, K. WEIHRAUCH, and X. ZHENG, *Weakly computable real numbers*, **Journal of Complexity**, vol. 16 (2000), no. 4, pp. 676–690.

[4] J. BĀRZDIŅŠ, *Complexity of programs to determine whether natural numbers not greater than n belong to a recursively enumerable set*, **Soviet Mathematics Doklady**, vol. 9 (1968), pp. 1251–1254.

[5] O. BELEGRADEK, *On algebraically closed groups*, **Algebra i Logika**, vol. 13 (1974), no. 3, pp. 813–816.

[6] C. CALUDE, **Information theory and randomness, an algorithmic perspective**, Springer-Verlag, Berlin, 1994.

[7] C. CALUDE, R. COLES, P. HERTLING, and B. KHOUSSAINOV, *Degree-theoretic aspects of computably enumerable reals*, **Models and computability** (Cooper and Truss, editors), Cambridge University Press, 1999.

[8] C. CALUDE, P. HERTLING, B. KHOUSSAINOV, and Y. WANG, *Recursively enumerable reals and Chaitin's $\omega$ number*, **Stacs '98**, Springer Lecture Notes in Computer Science, vol. 1373, 1998, pp. 596–606.

[9] C. CALUDE and A. NIES, *Chaitin's $\Omega$ numbers and strong reducibilities*, **Journal of Universal Computer Science**, vol. 3 (1997), pp. 1162–1166.

[10] G. CHAITIN, *A theory of program size formally identical to information theory*, **Journal of the Association for Computing Machinery**, vol. 22 (1975), pp. 329–340.

[11] ———, *Information-theoretical characterizations of recursive infinite strings*, **Theoretical Computer Science**, vol. 2 (1976), pp. 45–48.

[12] R. DOWNEY, *Computability, definability, and algebraic structures*, to appear, Proceedings of the 7th Asian Logic Conference, Taiwan.

[13] ———, *On the universal splitting property*, **Mathematical Logic Quarterly**, vol. 43 (1997), pp. 311–320.

[14] ———, *Some recent progress in algorithmic randomness*, **Proceedings, Mathematical Foundations of Computer Science**, 2004.

[15] R. Downey, D. Ding, S. P. Tung, Y. H. Qiu, M. Yasuugi, and G. Wu (editors), **Proceedings of the 7th and 8th Asian logic conferences**, World Scientific, 2003.

[16] R. DOWNEY and E. GRIFFITHS, *On Schnorr randomness*, **The Journal of Symbolic Logic**, vol. 62 (2004), pp. 533–554.

[17] R. DOWNEY, E. GRIFFITHS, and D. HIRSCHFELDT, *Submartingale reducibility and the calibration of randomness*, in preparation.

[18] R. DOWNEY, E. GRIFFITHS, and S. REID, *Kurtz randomness*, **Theoretical Computer Science**, (to appear).

[19] R. DOWNEY and D. HIRSCHFELDT, Walter De Gruyter, Berlin and New York, 2001, (Short courses in complexity from the New Zealand Mathematical Research Institute summer 2000 meeting, Kaikoura).

[20] ———, *Algorithmic randomness and complexity*, Monographs in Computer Science, Springer-Verlag, to appear.

[21] R. DOWNEY, D. HIRSCHFELDT, and G. LAFORTE, *Undecidability of Solovay and other degree structures for c.e. reals*, in preparation.

[22] ———, *Randomness and reducibility*, **Journal of Computing and System Sciences**, vol. 68 (2004), pp. 96–114, extended abstract in the **Proceedings of Mathematical Foundations of Computer Science 2001**, (J. Sgall, A. Pultr, and P. Kolman, editors), Lecture Notes in Computer Science 2136, Springer, 2001, pp. 316–327.

[23] R. DOWNEY, D. HIRSCHFELDT, and A. NIES, *Randomness, computability and density*, **SIAM Journal on Computing**, vol. 31 (2002), pp. 1169–1183, (Extended abstract in STACS'01).

[24] R. DOWNEY, D. HIRSCHFELDT, A. NIES, and F. STEPHAN, *Trivial reals*, **Proceedings of the 7th and 8th Asian logic conferences** (R. Downey, D. Ding, S. Tung, Y. Qiu, and M. Yasugi, editors), World Scientific, 2003, pp. 103–131.

[25] R. DOWNEY, D. HIRSCHFELDT, A. NIES, and S. TERWIJN, *Calibrating randomness*, **The Bulletin of Symbolic Logic**, (to appear).

[26] R. DOWNEY, C. JOCKUSCH, and M. STOB, *Array nonrecursive sets and multiple permitting arguments*, **Recursion theory week** (Ambos-Spies, Muller, and Sacks, editors), Lecture Notes in Mathematics, vol. 1432, 1990, pp. 141–174.

[27] R. DOWNEY and G. LAFORTE, *Presentations of computably enumerable reals*, **Theoretical Computer Science**, (to appear).

[28] R. DOWNEY and M. STOB, *Splitting theorems in recursion theory*, **Annals of Pure and Applied Logic**, vol. 65 (1993), no. 1, pp. 1–106.

[29] R. DOWNEY and S. TERWIJN, *Presentations of computably enumerable reals and ideals*, **Mathematical Logic Quarterly**, vol. 48 (2002), no. 1, pp. 29–40.

[30] R. G. DOWNEY, G. LAFORTE, and A. NIES, *Enumerable sets and quasi-reducibility*, **Annals of Pure and Applied Logic**, vol. 95 (1998), pp. 1–35.

[31] R. G. DOWNEY and J. B. REMMEL, *Classification of degree classes associated with r.e. subspaces*, **Annals of Pure and Applied Logic**, vol. 42 (1989), pp. 105–125.

[32] L. FORTNOW, *Kolmogorov complexity*, in [19], pp. 73–86.

[33] D. HIRSCHFELDT, personal communication, February 2001.

[34] CHUN-KUEN HO, *Relatively recursive reals and real functions*, **Theoretical Computer Science**, vol. 219 (1999), pp. 99–120.

[35] S. ISHMUKHAMETOV, *Weak recursive degrees and a problem of Spector*, **Recursion theory and complexity** (Arslanov and Lempp, editors), de Gruyter, Berlin, 1999, pp. 81–88.

[36] C. G. JOCKUSCH JR., *Fine degrees of word problems in cancellation semigroups*, **Zeitschrift für mathematische Logik und Grundlagen der Mathematik**, vol. 26 (1980), pp. 93–95.

[37] S. KAUTZ, *Degrees of random sets*, Ph.D. thesis, Cornell, 1991.

[38] KER-I KO, *On the continued fraction representation of computable real numbers*, **Theoretical Computer Science, A**, vol. 47 (1986), pp. 299–313.

[39] ———, *Complexity of real functions*, Birkhäuser, Berlin, 1991.

[40] KER-I KO and H. FRIEDMAN, *On the computational complexity of real functions*, **Theoretical Computer Science**, vol. 20 (1982), pp. 323–352.

[41] A. KUČERA, *On the use of diagonally nonrecursive functions*, **Logic colloquium, '87** (Amsterdam), North-Holland, 1989, pp. 219–239.

[42] ———, *On relative randomness*, **Annals of Pure and Applied Logic**, vol. 63 (1993), pp. 61–67.

[43] A. KUČERA and T. SLAMAN, *Randomness and recursive enumerability*, **SIAM Journal on Computing**, vol. 31 (2001), pp. 199–211.

[44] A. KUČERA and S. TERWIJN, *Lowness for the class of random sets*, **The Journal of Symbolic Logic**, vol. 64 (1999), pp. 1396–1402.

[45] M. KUMMER, *Kolmogorov complexity and instance complexity of recursively enumerable sets*, **SIAM Journal on Computing**, vol. 25 (1996), pp. 1123–1143.

[46] S. KURTZ, *Randomness and genericity in the degrees of unsolvability*, Ph.D. thesis, University of Illinois at Urbana.

[47] L. LEVIN, *Measures of complexity of finite objects (axiomatic description)*, **Soviet Mathematics Dolkady**, vol. 17 (1976), pp. 552–526.

[48] MING LI and P. VITANYI, *Kolmogorov complexity and its applications*, Springer-Verlag, 1993.

[49] D. LOVELAND, *A variant of the Kolmogorov concept of complexity*, **Information and Control**, vol. 15 (1969), pp. 510–526.

[50] A. MACINTYRE, *Omitting quantifier free types in generic structures*, **The Journal of Symbolic Logic**, vol. 37 (1972), pp. 512–520.

[51] P. MARTIN-LÖF, *The definition of random sequences*, **Information and Control**, vol. 9 (1966), pp. 602–619.

[52] M. POUR-EL, *From axiomatics to intrinsic characterization, some open problems in computable analysis*, **Theoretical Computer Science, A**, (to appear).

[53] M. POUR-EL and I. RICHARDS, *Computability in analysis and physics*, Springer-Verlag, Berlin, 1989.

[54] J. RASONNIER, *A mathematical proof of S. Shelah's theorem on the measure problem and related results*, **Israel Journal of Mathematics**, vol. 48 (1984), pp. 48–56.

[55] H. RICE, *Recursive real numbers*, **Proceedings of the American Mathematical Society**, vol. 5 (1954), pp. 784–791.

[56] C. P. SCHNORR, *A unified approach to the definition of a random sequence*, **Mathematical Systems Theory**, vol. 5 (1971), pp. 246–258.

[57] R. SOARE, *Cohesive sets and recursively enumerable Dedekind cuts*, **Pacific Journal of Mathematics**, vol. 31 (1969), no. 1, pp. 215–231.

[58] ———, *Recursion theory and Dedekind cuts*, **Transactions of the American Mathematical Society**, vol. 140 (1969), pp. 271–294.

[59] ———, *Recursively enumerable sets and degrees*, Springer, Berlin, 1987.

[60] R. SOLOMONOFF, *A formal theory of inductive inference, part 1 and part 2*, **Information and Control**, vol. 7 (1964), pp. 224–254.

[61] R. SOLOVAY, *Draft of paper (or series of papers) on Chaitin's work*, unpublished notes, May 1975, 215 pages.

[62] F. STEPHAN, personal communication.

[63] S. TERWIJN and D. ZAMBELLA, *Algorithmic randomness and lowness*, **The Journal of Symbolic Logic**, vol. 66 (2001), pp. 1199–1205.

[64] A. TURING, *On computable numbers with an application to the Entscheidungsproblem*, **Proceedings of the London Mathematical Society**, vol. 42 (1936), pp. 230–265, correction in Proceedings of the London Mathematical Society, 43 (1937), 544–546.

[65] M VAN LAMBALGEN, *Random sequences*, Ph.D. thesis, University of Amsterdam, 1987.

[66] K. WEIHRAUCH, *Computability*, Springer-Verlag, Berlin, 1987.

[67] K. WEIHRAUCH and X. ZHENG, *Arithmetical hierarchy of real numbers*, **Mathematical foundations of computer science 1999 (Sklarska Poreba, Poland)**, September 1999, pp. 23–33.

[68] GOUHUA WU, *Presentations of computable enumerable reals and the initial segment of computable enumerable degrees*, **Proceedings of COCOON' 01, Guilin, China**, 2001.

[69] M. ZIEGLER, *Algebraisch abgeschlossene Gruppen*, **Word problems II, the Oxford book** (Adian, Boone, and Higman, editors), North-Holland, 1980, pp. 449–576.

SCHOOL OF MATHEMATICAL AND COMPUTING SCIENCES
VICTORIA UNIVERSITY
PO BOX 600, WELLINGTON, NEW ZEALAND
*E-mail*: rod.downey@vuw.ac.nz

Index for *Some computability-theoretic aspects of reals and randomness*

# WEAK FRAGMENTS OF PEANO ARITHMETIC

PAOLA D'AQUINO

**§1. Introduction.** Our basic structures are discretely ordered rings, with corresponding first-order language $\mathcal{L}$ based on the symbols $+, \cdot, -, 0, 1, <$, with the usual interpretation and satisfying the further property $\forall x \neg (0 < x < 1)$. We will denote the theory of such rings by DOR. We will be mainly concerned with the nonnegative part of the ring where induction will be applied. There are various kinds of induction from a model-theoretic point of view, for example, first order and second order. There is a unique model of second-order induction, and this is $\mathbb{N}$.

The scheme of first-order induction is the following,

$$\forall \bar{y} \, (\theta \, (0, \bar{y}) \wedge \forall x \, (\theta \, (x, \bar{y}) \rightarrow \theta \, (x + 1, \bar{y})) \rightarrow \forall x \theta \, (x, \bar{y}))$$

where $\theta(x, \bar{y})$ runs through all formulas of $\mathcal{L}$. By varying the complexity of the formula $\theta$ we obtain the different subsystems of Peano Arithmetic (PA). I will be mainly interested in the theory $I\Delta_0$, where the induction is applied only to formulas with quantifiers bounded by terms of the language, and in theories related to $I\Delta_0$.

These systems are a natural setting for elementary number theory. One sees, simply by inspecting proofs, that all proofs in Hardy and Wright [20] can be done in PA,[1] but if we weaken the induction scheme some problems arise. The interest in studying classical results in weak systems is motivated by the following two facts:

(1) proving something in a weaker system often gives more constructive information. For example, Buss identified a subsystem of PA whose provably total functions are in the polynomial time hierarchy (see [8]);

---

The author thanks all the logicians at Notre Dame for their hospitality and support. Special thanks to J. Knight and P. Cholak for all the help they have given me when I was writing these notes and for their patience.

[1] In fact we claim that all of the results in Hardy and Wright [20] can be done in $I\Delta_0 + \exp$. For example, proof of the local version of Chebyshev's result is in Section 7. Another such example can be found in [10] where the authors show that the prime number theorem can be done in $I\Delta_0 + \exp$. The work in [45] provides futher justification.

**The Notre Dame Lectures**
Edited by P. Cholak
Lecture Notes in Logic, 18
© 2005, Association for Symbolic Logic

(2) weak fragments of arithmetic have connections with hard core complexity theory (where number theory has an increasingly important role, in primality testing, coding theory, and security communication).

All models I work with have a canonical part looking exactly like $\mathbb{Z}$, and, in addition (except for the standard model $\mathbb{Z}$), a nonstandard part consisting of *infinite* elements.

§2. **Open induction.** Let IOpen denote the subsystem of PA where induction is applied to formulas with no quantifiers (open formulas). There are algebraic methods for constructing models of IOpen, in contrast to what happens for all other fragments. Its models have been completely characterized in purely algebraic terms by Shepherdson in 1964 (see [43]) in the following way:

$\mathcal{M}$ is a model of IOpen iff $\mathcal{M}$ is a discretely ordered ring satisfying the condition

$$\text{for all } r \in RC(\mathcal{M}) \text{ there is } a \in \mathcal{M} \text{ such that } |a - r| < 1 \qquad (1)$$

where $RC(\mathcal{M})$ is the real closure of $\mathcal{M}$. We will denote the fraction field of $\mathcal{M}$ by $Q(\mathcal{M})$. Note that property (1) is first order. The elements of the real closure of $\mathcal{M}$ are algebraic over $\mathcal{M}$, hence roots of polynomials with coefficients in $\mathcal{M}$. Moreover, we can identify the $k$th root of a polynomial since we are in an ordered field. The proof of the Shepherdson result uses the elimination of quantifiers for real closed fields and gives the following characterization of models of IOpen (see [31]).

COROLLARY 2.1. *Let $\mathcal{M}$ be a discretely ordered ring. The following are equivalent*:

(1) $\mathcal{M} \models$ IOpen;

(2) *every quantifier-free definable subset of $\mathcal{M}$ (with parameters) is a finite union of intervals in $\mathcal{M}$*;

(3) *for every $r$ in $Q(\mathcal{M})$ there exists $a$ in $\mathcal{M}$ such that $|a - r| < 1$ and $Q(\mathcal{M})$ is dense in $RC(\mathcal{M})$*.

From Shepherdson's axiomatization of Open Induction it follows that for any $\alpha \in RC(\mathcal{M})$ we can define in the usual way the integer part of $\alpha$

$$[\alpha] = z \text{ iff } z \leq \alpha < z + 1 \qquad (2)$$

and $z \in \mathcal{M}$. We can now do Euclidean division of any given $x, y \in \mathcal{M}$, $y \neq 0$. Consider $x/y$ in $Q(\mathcal{M})$, and let $z = [x/y]$. Hence, $zy \leq x < y(z + 1)$, which is equivalent to saying that $x \in [zy, zy + y)$, i.e., $x = zy + r$ for some $r$ with $0 \leq r < y$. But we cannot carry on the Euclidean algorithm for constructing the g.c.d. of $x$ and $y$. If we could apply the Euclidean algorithm we would show that models of IOpen are Bezout rings, and in any Bezout ring the notions of irreducible and prime coincide. But Macintyre and Marker in [26] constructed a model of Open Induction where an irreducible element is not prime. Notice that if we work in $I\Delta_0$ we can prove by an easy $\Delta_0$-induction that

$$\forall a, b \exists d \leq a \exists x < b \exists y < a (d \mid a \wedge d \mid b \wedge d = ax + yb) \qquad (3)$$

where $z \mid w$ stands for the formula $\exists t \leq w(w = zt)$. It follows that any two elements in any model of $I\Delta_0$ have a greatest common divisor.

It is possible to construct very pathological models of IOpen by purely algebraic methods. For example, Shepherdson constructed a model of IOpen where the equation $x^2 = 2y^2$ has a nontrivial solution, so IOpen does not prove the irrationality of $\sqrt{2}$. We now sketch the construction of the Shepherdson model, which is obtained as a subring of the formal power series $\mathbb{R}((t^{\mathbb{Q}}))$. Let $\mathbb{Z}((t))$ denote the ring

$$\left\{ n_0 + \sum_{i=1}^{m} r_i t^{-\gamma_i} : n_0 \in \mathbb{Z}, r_i \in \mathbb{R}, \gamma_i \in \mathbb{Q}, 0 \leq \gamma_1 < \gamma_2 < \cdots < \gamma_m \right\}$$

with the order defined by

$$n_0 + \sum_{i=1}^{m} r_i t^{-\gamma_i} > 0 \text{ if } m \geq 1 \text{ and } r_m > 0, \text{ or if } n_0 > 0 \text{ otherwise.}$$

(Notice that $t$ has to be thought of as *small*, e.g., $t^{-1} = x$, where $x$ is an infinite positive element.) The positive part of $\mathbb{Z}((t))$ satisfies induction on open formulas, so it is a model of IOpen. This model has many properties which do not hold in stronger subsystems of PA. It is well known that nonstandard models of PA and of $I\Delta_0$ are subject to the Tennenbaum phenomenon, i.e., the operations of $+$ and $\cdot$ are not recursive. On the contrary, $\mathbb{Z}((t))$ is recursive. Moreover, $(\sqrt{2}t^{-1}, t^{-1})$ is a nontrivial solution of $x^2 = 2y^2$, and $(t^{-1}, t^{-1}, \sqrt[n]{2}t^{-1})$ is a nontrivial solution of $x^n + y^n = z^n$, which implies that Fermat's Last Theorem is not provable in IOpen. Wilkie noticed that the only irreducible elements of $\mathbb{Z}((t))$ are the standard ones (for the proof see [48]). The element $1 + t^{-1}$ is not divisible by any prime $p \in \mathbb{N}$ since any $p$ divides $t^{-1}$.

Berarducci and Otero in [7] constructed a recursive nonstandard model of IOpen with an unbounded set of infinite prime elements. It turns out that their model is also normal, i.e., it is integrally closed in its fraction field. Normality is a first-order property; we need to express that any element of the fraction field of $\mathcal{M}$ which is a root of a monic polynomial with coefficients in $\mathcal{M}$ is in fact an element of $\mathcal{M}$. The formalization of such property is the following formula $N$

$$\forall \bar{z} \forall x, y \left( x, y \neq 0 \wedge x^n + z_1 x^{n-1} y + \cdots \right.$$
$$\left. + z_{n-1} x y^{n-1} + z_n y^n = 0 \rightarrow \exists w (x = wy) \right)$$

where $n \in \mathbb{N}$. In general, models of IOpen are not normal: normality follows from unique factorization and, for example, this does not hold in the Shepherdson model. Van den Dries was the first to consider normal models of Open Induction in [46]. Such models avoid certain pathologies, for example, it is possible to show in IOpen $+N$ that $\sqrt{2}$ is irrational. From the result of Berarducci and Otero it follows that IOpen $+N$ is not subject to the Tennenbaum phenomenon. For other results on Normal Open Induction see [32], [33], [34].

**§3. Bounded induction.** We now consider the subsystem of Peano Arithmetic $I\Delta_0$ where induction is applied to formulas with bounded quantifiers. Constructing models of $I\Delta_0$ is not easy, the $\Delta_0$-induction puts too many constrains in developing algebraic constructions. As already remarked, models of $I\Delta_0$ are Bezout rings, hence the notions of irreducible element and prime element coincide.

There are very few methods for constructing models of $I\Delta_0$.

LEMMA 3.1. *Let $\mathcal{M}$ be a model of $I\Delta_0$ and let $I$ be an initial segment of $\mathcal{M}$. If $I$ is closed under $+$ and $\cdot$ then $I$ is a model of $I\Delta_0$.*

PROOF. Suppose that $I \models \theta(0) \wedge \forall x(\theta(x) \rightarrow \theta(x+1))$ and $I \models \neg\theta(a)$, for $\theta \in \Delta_0$ and $a \in I$. $\Delta_0$-formulas are preserved from a model to any initial segment and vice versa. The set $X = \{x : \mathcal{M} \models \neg\theta(x)\}$ is $\Delta_0$-definable and not empty. Let $a$ be the minimum of $X$, hence $\mathcal{M} \models \theta(a-1)$, which implies $I \models \theta(a-1)$. From the hypothesis, it follows that $I \models \theta(a)$, and so also $\mathcal{M} \models \theta(a)$, which gives a contradiction. The same proof works with parameters.                                                                    ⊣

If $\mathcal{M}$ is a model of $I\Delta_0$ and $a \in \mathcal{M}$, then the set $BP[a] = \{x \in \mathcal{M} : x < a^n, n \in \mathbb{N}\}$ is an initial segment closed under $+$ and $\cdot$, and so it is a model of $I\Delta_0$.

Parikh in 1971 (see [35]) focused attention on $I\Delta_0$ by proving the following result.

THEOREM 3.2. *Let $\theta(\bar{x}, y) \in \Delta_0$. Suppose that $I\Delta_0 \vdash \forall\bar{x}\exists y\theta(\bar{x}, y)$. Then $I\Delta_0 \vdash \forall\bar{x}\exists y < \tau(\bar{x})\theta(\bar{x}, y)$, where $\tau(\bar{x})$ is a term of $\mathcal{L}$.*

PROOF. For simplicity we restrict to the case where $\bar{x}$ is a single variable. The proof in the general case follows the same lines. Suppose that $I\Delta_0 \vdash \forall x\exists y\theta(x, y)$ and $I\Delta_0 \nvdash \forall x\exists y < \tau(x)(\theta(x, y))$, for each term $\tau(x)$. Add a new constant symbol $c$ to the language. By a simple compactness argument, the theory $T = I\Delta_0 \cup \{\forall y < \tau(c)\neg\theta(c, y) : \tau(c)$ term of the language $\mathcal{L}_c\}$ has a model $\mathcal{M}$. Let $a = c^{\mathcal{M}}$ and consider the initial segment of $\mathcal{M}$ generated by the standard powers of $a$, i.e., $BP[a] = \{x \in \mathcal{M} : x < a^n, n \in \mathbb{N}\}$. Since $\Delta_0$-formulas are preserved under end-extensions and initial segments, from $\mathcal{M} \models \forall y < \tau(a)\neg\theta(a, y)$ for all terms $\tau$ it follows that $BP[a] \models \forall y < \tau(a)\neg\theta(a, y)$ for all terms $\tau$. We get then a contradiction, since $BP[a]$ is a model of $I\Delta_0$ and $I\Delta_0 \vdash \forall x\exists y\theta(x, y)$.                                                ⊣

When working in weak fragments of PA such as $I\Delta_0$, we have to deal with two kinds of problems:

(i) we need to express concepts in a $\Delta_0$-way in order to extend classical notions to any nonstandard model;

(ii) functions with $\Delta_0$-graphs and exponential growth are not provably total in $I\Delta_0$.

If on one hand, Parikh's result showed a main limitation of $I\Delta_0$, on the other hand, it makes $I\Delta_0$ interesting from a number theoretical point of view. New proofs of classical results are produced and more information on the computational aspect of old proofs is obtained.

For example, the classical result on the cofinality of primes is proved using the function $\prod_{p\leq x} p = y$, with $p$ prime, which is definable by the $\Delta_0$-formula

$$\forall p \leq x(Pr(p) \rightarrow p \mid y \wedge p^2 \nmid x) \wedge \forall p \leq y(Pr(p) \wedge p > x \rightarrow p \nmid y)$$

where $p \mid y$ stands for $\exists q \leq y(y = pq)$ and $Pr(p)$ stands for $\forall z \leq p(z \mid p \leftrightarrow z = 1 \vee z = p)$. Such a function is of exponential growth (Chebyshev's theorem); hence, it is not provably total in $I\Delta_0$. So, the usual proof of cofinality of primes is not formalizable in $I\Delta_0$. It is one of the main open problems in this area if $I\Delta_0$ proves cofinality of primes.

QUESTION 3.3 (Wilkie). *Does $I\Delta_0 \vdash \forall x \exists p(Pr(p) \wedge p > x)$?*

If we add an axiom which guarantees the totality of the exponential function, i.e.,

$$\exp: \forall x \exists y(2^x = y)$$

(where $2^x = y$ stands for a $\Delta_0$-formula defining exponentiation which we will discuss in Section 4) then the system $I\Delta_0 + \exp$ is strong enough to reproduce almost all of elementary number theory. In particular, $I\Delta_0 + \exp \vdash$ cofinality of primes. But working in a system like $I\Delta_0 + \exp$ does not give constructive information on proofs; we simply adapt the classical arguments getting only iterated exponential bounds.

For the theory IOpen, from the result of Macintyre and Marker, it follows that there are models where the only primes are the standard ones.

As already recalled, models of $I\Delta_0$ are Bezout rings, and so the notions of prime and irreducible coincide. This implies that *prime* is $\Delta_0$-definable. We next show that models of $I\Delta_0$ are unique factorization domains in the sense that we can extend the notion of the factorization in powers of primes for any element in a nonstandard model of $I\Delta_0$.

First of all we prove the following two lemmas concerning general $\Delta_0$-definable sets in a model $\mathcal{M}$ of $I\Delta_0$.

LEMMA 3.4. *Let $A$ be a bounded $\Delta_0$-definable subset of $\mathcal{M}$. Then $A$ has a maximum element.*

PROOF. Let $\beta$ be an upper bound of $A$, and consider the set $X = \{y \leq \beta: \exists t \in A(y < t)\}$. Then $X$ is clearly $\Delta_0$-definable, and so is $\mathcal{M} - X = \{y: y > \beta\} \cup \{y: y \leq \beta \wedge y > t \forall t \in A\}$. Let $x_0 = \min(\mathcal{M} - X)$. Then $x_0 - 1 \in X$, and so there exists $t \in A$ such that $x_0 - 1 \leq t$. Necessarily, $x_0 - 1 = t$, which implies that $x_0 - 1 = \max(A)$.                    ⊣

LEMMA 3.5. *Let $A$ be a bounded $\Delta_0$-definable subset of $\mathcal{M}$. If $\mathcal{M}$ contains an element $m$ divisible by all $a \in A$, then there exists $\mu \in \mathcal{M}$ which is divisible by all $a \in A$ and is minimal with respect to this property; i.e., if $x \in \mathcal{M}$ is divisible by all elements of $A$, then $\mu$ divides $x$.*

PROOF. Let $\phi(x)$ be the $\Delta_0$-formula defining the set $A$. Consider the following $\Delta_0$-set

$$D = \{b: \forall a \leq m(\phi(a) \rightarrow a \mid b)\}.$$

Note that $D$ is nonempty since $m \in D$. Let $\mu$ be the minimum of $D$. If $x$ is divisible by all $a \in A$ and not by $\mu$, then $x = \mu q + r$ for some $q, r \in \mathcal{M}$ with $0 \leq r < \mu$. Hence, $r$ is divisible by all $a \in A$, and from the minimality of $\mu$, it follows that $r = 0$.                    ⊣

We will refer to $\mu$ as the *least common multiple of $A$*.

By an easy $\Delta_0$-induction, we can prove that any element is divisible by a prime.

LEMMA 3.6. $I\Delta_0 \vdash \forall x > 1 \exists p \leq x(Pr(p) \wedge p \mid x)$.

We will use the notation $\text{Pow}_p(x)$ to denote the formula $Pr(p) \wedge \forall y \leq x(y \mid x \rightarrow p \mid y)$ which says that $x$ is a power of a prime $p$.

LEMMA 3.7.

$$I\Delta_0 \vdash \forall x > 1 \forall p(Pr(p) \rightarrow \exists! y \leq x(\text{Pow}_p(y) \wedge y \mid x \wedge py \nmid x)).$$

PROOF. Let $x$ be an element of $\mathcal{M} \models I\Delta_0$ and $p$ a prime in $\mathcal{M}$. Consider the following $\Delta_0$-set

$$A_p^x = \{y: \text{Pow}_p(y) \wedge y \mid x\}.$$

Clearly, $A_p^x$ is bounded by $x$, and so by Lemma 3.4, it has a maximum element, which is the greatest power of $p$ that divides $x$.                    ⊣

$M \text{Pow}_p(x, y)$ stands for the $\Delta_0$-formula

$$\text{Pow}_p(y) \wedge y \mid x \wedge \forall z \leq x(\text{Pow}_p(z) \wedge z \mid x \rightarrow z \mid y).$$

Given $x \in \mathcal{M}$, consider the set

$$A_x = \{y: \exists p \leq x(Pr(p) \wedge M \text{Pow}_p(x, y)\}.$$

Clearly, $A_x$ is $\Delta_0$-definable, and there exists an element, namely $x$, divisible by all elements of $A$. Hence, there is the smallest element $\mu$ divisible by all elements of $A$, and it is trivial to show that $\mu$ coincides with $x$. So we have extended the notion of factorization in powers of primes for every element $x$ of any model of $I\Delta_0$ as the $\text{lcm}(A_x)$.

There is a version of the Chinese Remainder Theorem which is available in $I\Delta_0$.

THEOREM 3.8 (CRT). *Let $A$ be a bounded $\Delta_0$-definable set in $\mathcal{M}$. Let $f, r: A \longrightarrow \mathcal{M}$ be $\Delta_0$-definable functions such that $(f(a), f(b)) = 1$ for all $a, b \in A$ and $r(a) < f(a)$ for all $a \in A$. Suppose there is $w \in \mathcal{M}$ which is divisible by all elements of $f(A)$. Then there exists $u < \prod_{a \in A} f(a)$ such that $u \equiv r(a)(\mathrm{mod}\, f(a))$.*

PROOF. Consider the $\Delta_0$-formula

$$\psi(x, w) = \exists u \leq w$$

$$(u < \prod_{\substack{a \in A \\ a \leq x}} f(a) \wedge \forall t \leq x(t \in A \rightarrow \mathrm{Rem}(u, f(t)) = r(t)),$$

where $\mathrm{Rem}(u, x)$ stands for the remainder of $u$ divided by $x$. We want to show that $\mathcal{M} \models \forall x(x \in A \rightarrow \psi(x, w))$. Use $\Delta_0$-induction: let $u^* < \prod_{a \in A, a \leq x} f(a)$ and $u^* \equiv r(a)(\mathrm{mod}\, f(a))$ for all $a \leq x$. Now solve the system of two congruences

$$u \equiv u^*(\mathrm{mod} \prod_{\substack{a \leq x \\ a \in A}} f(a))$$

$$u \equiv r(x + 1)(\mathrm{mod}\, f(x + 1)).$$

This is solvable since the moduli are coprime. ⊣

Notice that $\prod_{a \in A} f(a)$ has to be interpreted as the $\mathrm{lcm}(f(A))$, which exists because of the hypothesis that for all $a \in A$ $f(a)$ divides $w$ for some $w$.

The Chinese Remainder Theorem can be used to give $\Delta_0$-definitions of the following two functions

$$p_n(x) = n\text{th prime divisor of } x \text{ (in order of magnitude)}$$
$$v(x) = \text{number of distinct prime divisors of } x.$$

LEMMA 3.9. *For any given $x \in \mathcal{M}$, there is $c \in \mathcal{M}$ such that*
*(i) $c \equiv 1(\mathrm{mod}\, p)$ if $p$ is the least prime dividing $x$;*
*(ii) if $p, q$ are prime divisors of $x$, $p < q$ and there is no prime divisor of $x$ between $p$ and $q$ then $c \equiv n(\mathrm{mod}\, p)$ implies $c \equiv n + 1(\mathrm{mod}\, q)$.*

PROOF. Let $A$ be the set of all primes that divide $x$. We can apply CRT since there is an element, $x$ itself, which is divisible by all primes in $A$. ⊣

So, if $n < p$ and $p$ divides $x$, then $p = p_n(x)$ if and only if $c \equiv n(\mathrm{mod}\, p)$. Also, $v(x) = k$ if and only if $p_k(x)$ is the largest prime divisor of $x$.

**§4. Exponentiation.** It was Gödel who first showed using the Chinese Remainder Theorem that the graph of the exponential function is definable in terms of $+$ and $\cdot$. In order to define the relation $x^y = z$, we need to code the sequence of values

$$1, x, x^2, x^3, \ldots, x^{y-1}, x^y.$$

Take $u, v$ such that $1 + v, 1 + 2v, \ldots, 1 + (1 + y)v$ are coprime and $u \equiv x^i \pmod{1 + (i + 1)v}, i = 1, \ldots, y$. Then $x^y = z$ is defined by the $\Sigma_1$-formula

$$\exists u, v \,(\mathrm{Rem}(u, 1 + v) = 1$$
$$\wedge\, \forall t < y \,(t > 0 \rightarrow \mathrm{Rem}(u, 1 + (t + 1)v) = x\,\mathrm{Rem}(u, 1 + tv)$$
$$\wedge\, \mathrm{Rem}(u, 1 + (y + 1)v) = z)). \quad (4)$$

We will denote such a formula by $E(x, y, z)$. The following estimates on $u$ and $v$ satisfying the conditions above hold:

$$u < \prod_{i=0}^{y}(1 + v(i + 1)) \text{ and } v < y!z.$$

Clearly, these are of exponential size, hence the existential quantifiers in $E(x, y, z)$ cannot be bounded by polynomials. In $I\Sigma_1$, the formula $E(x, y, z)$ satisfies the recursion laws of exponentiation and gives an *invariant* meaning to $x^y$ in every model of $I\Sigma_1$.

Using only $\Delta_0$-induction, we can only prove that the relation defined by $E(x, y, z)$ is functional and it is defined on an initial segment; i.e.,

$$I\Delta_0 \vdash \forall x > 0 \forall y, z, t(E(x, y, z) \wedge E(x, y, t) \rightarrow z = t)$$
$$I\Delta_0 \vdash \forall x > 0 \forall y, z(E(x, y + 1, z) \rightarrow \exists w < z(E(x, y, w) \wedge z = xw)).$$

While it is easily seen that $u, v$ coding a sequence of length $y + 2$ code also a sequence of length $y + 1$, there is no immediate way of getting $u, v$ coding the sequence $1, x, x^2, \ldots, x^{y+1}$ from $u', v'$ coding the sequence $1, x, x^2, \ldots, x^y$. This is a consequence of the fact that there is no guarantee on the existence of the $\mathrm{lcm}(1 + v, 1 + 2v, \ldots, 1 + (y + 2)v)$ from the $\mathrm{lcm}(1 + v, 1 + 2v, \ldots, 1 + (y + 1)v)$. This is the reason why we cannot prove in $I\Delta_0$ the recursion laws of exponentiation for $E(x, y, z)$.

It was Bennett [4] in 1962 who first found a $\Delta_0$-formula defining the graph of exponentiation in $\mathbb{N}$. Twenty years later, Paris (see [18]) found a $\Delta_0$-definition of exponentiation, for which the recursive properties of exponentiation can be proved in $I\Delta_0$, and which gives an *invariant* meaning to $x^y = z$ in any nonstandard model of $I\Delta_0$. Let $E_0(x, y, z)$ be such a formula.

THEOREM 4.1. $I\Delta_0 \vdash$
(1) $\forall x > 0 \forall y, z, t(E_0(x, y, z) \wedge E_0(x, y, t) \rightarrow z = t)$;
(2) $\forall x > 0 \forall y, z(E_0(x, y + 1, z) \rightarrow \exists w < z(E_0(x, y, w) \wedge z = xw))$;

(3) $E_0(x, 0, 1) \wedge \forall x > 0 \forall y, z(E_0(x, y, z) \rightarrow E_0(x, y+1, xz))$; *and*
(4) *for any $\Delta_0$-formula $\theta(x, y, z)$ for which $I\Delta_0$ proves the properties in* (1),
(2), *and* (3) *we have* $I\Delta_0 \vdash \forall x > 0 \forall y, z(E_0(x, y, z) \leftrightarrow \theta(x, y, z))$.

REMARK 4.2. Fix $x$ in a nonstandard model $\mathcal{M}$ of $I\Delta_0$, then both the
domain of definition $G_x$ of the Gödel formula and the domain of definition
$B_x$ of the formula $E_0$ are initial segments of $\mathcal{M}$. By an easy $\Delta_0$-induction, it
can be proved that $E(x, y, z) \longrightarrow E_0(x, y, z)$, which implies that $G_x \subseteq B_x$.
The other inclusion is still an open problem, and it is connected with the
problem of guaranteeing the existence of the lcm of a set with a nonstandard
number of elements, as we discussed above.

There is an alternative $\Delta_0$-definition of the graph of the exponential function
due to Pudlak [42] which can be proved equivalent to $E_0(x, y, z)$ over $I\Delta_0$. We
now give an idea on how the formula $E_0$ is constructed. For the full details
see [18].

STEP 1. We want to define $x^y = z$. Consider the dyadic expansion of
$y = a_0 2^s + a_1 2^{s-1} + \cdots + a_s 2^0$ with $a_i = 0, 1$ and $s = |y|$. We construct two
sequences of logarithmic length which approximate $y$ and $x^y$. Let

$$\eta_0 = a_0 2^0$$
$$\eta_1 = a_0 2^1 + a_1 2^0$$
$$\eta_2 = a_0 2^2 + a_1 2^1 + a_2 2^0$$
$$\vdots$$
$$\eta_{s-1} = a_0 2^{s-1} + \cdots + a_{s-1} 2^0$$
$$\eta_s = a_0 2^s + a_1 2^{s-1} + \cdots + a_s 2^0$$
$$= y$$

and

$$\xi_0 = x$$
$$\xi_{i+1} = \xi^2 \quad \text{if } a_{i+1} = 0$$
$$\xi_{i+1} = x\xi^2 \quad \text{if } a_{i+1} = 1$$

for $i < s$. The following relations hold $\eta_{i+1} - 2\eta_i = a_{i+1}$ and $\xi_i = x^{\eta_i}$ for
all $i < s$.

STEP 2. Using the Chinese Remainder Theorem, we now define in a $\Delta_0$-way
the relation "$u, v$ code the first $s + 1$ primes". The formula is the following.

$$R(u, v, s) = v < u \wedge (\mathrm{Rem}(u, 2) = 0)$$
$$\wedge \forall p, q < v(C(p, q) \wedge (\mathrm{Rem}(u, q) = \mathrm{Rem}(u, p) + 1)$$
$$\wedge \exists r(Pr(r) \wedge \mathrm{Rem}(u, s) = r)),$$

where $C(p, q)$ says that $p, q$ are consecutive primes, i.e.,

$$C(p, q) = p < q \land Pr(p) \land Pr(q) \land \forall x(p < x < q \rightarrow \neg Pr(x)).$$

The following property provable in $I\Delta_0$ guarantees that if $u, v$ code the first $s$ primes then we can identify the $i$th prime.

LEMMA 4.3.

$$I\Delta_0 \vdash \forall u, v, s(R(u, v, s) \rightarrow \forall i \leq s \exists! p < v(Pr(p) \land Rem(u, p) = i)).$$

The formula $p_i(u, v)$ denotes the unique prime $p$ less than $v$ which satisfies $Rem(u, p) = i$. The $i$th prime does not depend on $u, v$ coding the first $s$ primes, as the following lemma guarantees.

LEMMA 4.4.

$$I\Delta_0 \vdash \forall u, v, u', v', s(R(u, v, s) \land R(u', v', s)$$
$$\rightarrow \forall i \leq s(p_i(u, v) = p_i(u', v'))).$$

STEP 3. We require now the existence of $w$ which is the product of powers of the first $s + 1$ primes coded by the pair $(u, v)$. Let $w = \prod_{i=0}^{s} w_i$, where $w_i = p_i^{k_i}$ with $i = 0, \ldots, s$. We now use the Chinese Remainder Theorem in order to code the two pairs of sequences $\eta_0, \eta_1, \ldots, \eta_s$ and $w_0, w_1, \ldots, w_s$ and $\xi_0, \xi_1, \ldots, \xi_s$ and $w_0, w_1, \ldots, w_s$. Let $\eta$ and $\xi$ be such that

$$Rem(\eta, w_0) = \eta_0 \land Rem(\xi, w_0) = x \land Rem(\eta, w_s) = y \land Rem(\xi, w_s) = z$$

$$\land \forall i < s \, [(Rem(\eta, w_{i+1}) = 2 \, Rem(\eta, w_i) \land Rem(\xi, w_{i+1}) = Rem(\xi, w_i)^2)$$

$$\lor (Rem(\eta, w_{i+1}) = 2 \, Rem(\eta, w_i) + 1 \land Rem(\xi, w_{i+1}) = x \, Rem(\xi, w_i)^2)].$$

Paris obtained the bound $2z^4$ on the size of $u, v, s, w, \eta, \xi$. He proved that if there are $w, u, v, \eta, \xi$ satisfying the above properties then there are $w', \eta', \xi', u', v' < z^4$ satisfying the same relations. Because of the following inequalities it is enough to bound the size of $w$:

$$u, v < \operatorname{lcm}_{i \leq s}(p_i) \leq \operatorname{lcm}_{i \leq s}(w_i) = w \text{ and } \eta, \xi < \operatorname{lcm}_{i \leq s}(w_i) = w. \quad (5)$$

The following properties can be proved by $\Delta_0$-induction.

LEMMA 4.5. $I\Delta_0$ proves
(1) $\forall i \leq s(p_i \leq \xi_i)$,
(2) $\forall i \leq s(w_i \leq \xi_i^2)$,
(3) $\forall i \leq s(\prod_{j < i} w_i \leq \xi_i^2)$.

The size of $w$ is bounded by pushing down the factors, the $w_i$s, and using the Chinese Remainder Theorem every time a factor $w_i$ is reduced.

From the $\Delta_0$-definition of exponentiation, we get easily that the logarithm function is $\Delta_0$-definable:

$$y = [\log_x z] \text{ if and only if } x^y \leq z < x^{y+1}. \quad (6)$$

Moreover, logarithm is a total function. For any given $z$, consider the set $A = \{w \le z : w$ is a power of $x\}$. $A$ is $\Delta_0$-definable and is bounded by $z$; then by Lemma 3.4, $A$ has a maximum element, and this uniquely defines the logarithm of $z$ in basis $x$.

The $\Delta_0$-definition of exponentiation is also used in the $\Delta_0$-definition of the $p$-adic valuation of an element for fixed $p$.

$$v_p(a) = b \text{ if and only if } p^b \mid a \wedge p^{b+1} \nmid a. \tag{7}$$

### §5. McAloon's theorem. In [29] McAloon proved the following theorem.

THEOREM 5.1. *Any countable nonstandard model of $I\Delta_0$ has a nonstandard initial segment which is a model of* PA.

McAloon's proof uses diagonal indiscernibles and his argument can be extended also to uncountable models. He worked with countable models since he was principally interested in their complexity in terms of recursion theory. He also showed that the initial segment $I$ can be chosen recursively saturated. His main result uses Friedman's embedding theorem which depends on the countability of the model. We will give an alternative proof using saturation which avoids appealing to Friedman's theorem.

DEFINITION 5.2. Let $(A, <)$ be an ordered structure. A subset $X$ of $A$ is a set of *diagonal indiscernibles* for a set $C$ of formulas if

$$\text{for all } \phi(v_0, \ldots, v_n, w_0, \ldots, w_m) \in C$$

and

$$\text{for all } c < c_0 < \cdots < c_n \text{ and } c < d_0 < \cdots < d_n \text{ and all } t_0 < \cdots < t_m < c$$

we have

$$A \models \phi(c_0, \ldots, c_n, t_0, \ldots, t_m) \leftrightarrow \phi(d_0, \ldots, d_n, t_0, \ldots, t_m).$$

The importance of diagonal indiscernibles follows from the fact that from a suitable set of diagonal indiscernibles in a model of $I\Delta_0$ one can construct a model of PA. More precisely, let $\mathcal{M}$ be a model of $I\Delta_0$, and let $J$ be a nonempty set of diagonal indiscernibles for $\Delta_0$-formulas with the further property that if $c, d \in J$ and $c < d$ then $c^2 < d$. Then $B_J = \{x : x < c \text{ for some } c \in J\}$ is a model of PA. So, in order to prove McAloon's result it is enough to realize the following recursive type $\tau(t_0, t_1, \ldots)$ of $\Delta_0$-formulas in countably many unknowns.

$$\{t_i < t_j : i < j\} \cup \{t_i^2 < t_j : i < j\}$$

$$\cup \{\forall w_0, \ldots, w_m < t_j \phi(t_{j_0}, \ldots, t_{j_n}, w_0, \ldots, w_m)$$

$$\leftrightarrow \phi(t_{i_0}, \ldots, t_{i_n}, w_0, \ldots, w_m):$$

$$\phi \in \Delta_0, j < j_0 < \ldots, j_n, j < i_0 < \ldots, i_n\}.$$

There is a standard device for obtaining a type in a single variable $t$ by replacing $t_i$ by $v_{p_i}(t)$, where $v_{p_i}(t)$ denotes the $p_i$-adic valuation of $t$.

A model of $I\Delta_0$ is not in general $\Delta_0$-recursively saturated, as a model of PA is not in general recursively saturated. The study of recursive saturation for models of $I\Delta_0$ is related to the still open problem of the collapse of the $\Delta_0$-hierarchy. For more detail see [39] and [47]. The following is known about recursive saturation for a model of $I\Delta_0$:

(1) in a model of PA any initial segment determined by an element $[0, a]$ with the inherited relational structure is recursively saturated; on the contrary, there is an example due to Paris and Dimitracopoulos of an initial segment $[0, a]$ of a model of $I\Delta_0$ which is not recursively saturated (see [39]).

(2) Lessan proved that if $2^{b^n}$ is defined for all $n \in \mathbb{N}$ and there is $a > 2^{b^n}$ for all $n \in \mathbb{N}$, then $[0, b]$ is recursively saturated (see [24]). This follows from the fact that under the hypothesis above, there is a formula $\Gamma(a, y, z)$ which defines satisfaction for all formulas over the structure $[0, b]$ in the structure $[0, a]$. The formula $\Gamma$ is $\Delta_0$ since all quantifiers are bounded by $a$.

It turns out that in any model $\mathcal{M}$ of $I\Delta_0$ there are always elements $a, b$ satisfying the properties above. Take $x \in \mathcal{M}$ such that $2^x$ is defined. The elements $b = \log_2 x$ and $a = 2^x$ satisfy the conditions above. So we can always find an initial segment $[0, b]$ which is recursively saturated. We now try to realize the type $\tau$ below such a $b$. By a result of Paris and Harrington (see [38]), the type $\tau$ is finitely satisfiable in $\mathbb{N}$, and since $[0, b]$ is recursively saturated, it is realized below $b$. So we get a set of diagonal indiscernibles in $\mathcal{M}$ which generates the nonstandard model of PA.

Notice that the proof of McAloon's result using saturation does not need any restriction on the cardinality of the model.

We now consider the following more general problem.

Given $a \in \mathcal{M} \models I\Delta_0$ when is there an initial segment $I$ of $\mathcal{M}$ such that $a \in I$ and $I \models$ PA?

As in the proof of McAloon's result, it will be sufficient to construct a set $J$ of diagonal indiscernibles such that $a < J$. This leads to the problem of satisfying the type

$$\tau^*(a, t_0, t_1, \dots) = \tau(t_0, t_1, \dots) \cup \{a < t_i : i \in \mathbb{N}\}.$$

The answer to this question is given in terms of the functions of the Wainer hierarchy (which presupposes a system of ordinal notation up to $\varepsilon_0$).

DEFINITION 5.3. For $\alpha$ an ordinal and $n$ a natural number, let $\omega_n^\alpha$ be defined inductively by

$$\omega_0^\alpha = \alpha$$
$$\omega_{n+1}^\alpha = \omega^{\omega_n^\alpha}.$$

Also let,

$$\omega_n = \omega_n^1,$$

and

$$\varepsilon_0 = \sup_{n \in \mathbb{N}} \omega_n.$$

The functions of the Wainer hierarchy are defined as follows.

DEFINITION 5.4. For $\alpha < \varepsilon_0$,

$$F_0(x) = x + 1;$$
$$F_{\alpha+1}(x) = F_{\alpha+1}^{(x+1)}(x);$$
$$F_\alpha(x) = F_{\{\alpha\}(x)}(x);$$

if $\alpha$ is a limit ordinal and $\{\alpha\}(x)$ denotes the $x$th element of the fundamental sequence of $\alpha$.

It is well known that the functions $F_\alpha$, $\alpha < \varepsilon_0$ are provably total in PA. The first $\omega$-level is the Ackermann hierarchy and $F_\omega$ is the Ackermann function. In [12] a $\Delta_0$-definition of the graphs of the functions of the Ackermann hierarchy is given, and the recursion properties of these functions are proved in $I\Delta_0$. A much more general and complete result has been obtained by Sommer [44]. He first extended the notion of ordinal $< \varepsilon_0$ to any model of $I\Delta_0$ in such a way that many of the basic facts about ordinals $< \varepsilon_0$ can be proved in $I\Delta_0$. He then found a $\Delta_0$-definition of the graphs of the functions $F_\alpha$ for $\alpha < \varepsilon_0$, and he proved the recursive properties of the $F_\alpha$s in $I\Delta_0$. So in any model of $I\Delta_0$ we can handle the functions of the Wainer hierarchy, even if they are only partial functions, but where they are defined they behave *well*. Sommer used them in order to prove the following result.

THEOREM 5.5. *Let $\mathcal{M}$ be a model of $I\Delta_0 + \exp$ and $a \in \mathcal{M}$. Fix $n \in \mathbb{N}$. A necessary and sufficient condition for $a$ to belong to a nonstandard initial segment $I \models I\Sigma_n$ is that $F_\alpha(a)$ is defined for all $\alpha < \omega_n$, and for some $b \in \mathcal{M}$, $F_\alpha(a) < b$ for all $\alpha < \omega_n$.*

REMARK 5.6. The conditions expressed in Sommer's result can be shown to be necessary and sufficient also in the case of $I\Delta_0$ instead of $I\Delta_0 + \exp$ (see [12]).

The answer to our original problem is contained in the following theorem.

THEOREM 5.7. *Let $\mathcal{M} \models I\Delta_0$ and $a \in \mathcal{M}$. The following are equivalent:*
*(1) there exists an initial segment $J$ of $\mathcal{M}$ such that $a \in J$ and $J \models PA$;*
*(2) $F_\alpha(a)$ is defined for all $\alpha < \varepsilon_0$ and there exists $b \in \mathcal{M}$ such that $F_\alpha(a) < b$ for all $\alpha < \varepsilon_0$.*

Theorem 5.5 is used in order to guarantee the finite satisfiability of type $\tau^*(a, t_0, t_1, \dots)$. The details of the proof can be found in [12].

**§6. Pigeonhole principle.** Woods in [51] focused attention on the combinatorial principle of the pigeonhole and showed that if it is added to $I\Delta_0$, the resulting theory is strong enough to prove the cofinality of primes. He discovered a new proof of this classical result based on some ideas of Sylvester. Woods' result shows that sometimes we can avoid functions of exponential growth and replace them by combinatorial principles. He considered the following version of the pigeonhole principle:

$$\forall \bar{w} \forall z (\forall x < z + 1 \exists y < z \theta(x, y, \bar{w}) \rightarrow \exists x_1, x_2 < z + 1 \exists y < z$$
$$(x_1 \neq x_2 \wedge \theta(x_1, y, \bar{w}) \wedge \theta(x_2, y, \bar{w}))) \quad (\Delta_0\text{-PHP})$$

where $\theta(x, y, \bar{w}) \in \Delta_0$.

THEOREM 6.1 (Woods). $I\Delta_0 + \Delta_0\text{-PHP} \vdash \forall x \exists p (Pr(p) \wedge p > x)$.

It is clear that $I\Delta_0 + \exp \vdash \Delta_0\text{-PHP}$, while the opposite is not true. $\Delta_0\text{-PHP}$ is a $\prod_1$-statement; hence, it is preserved under initial segment. Let $\mathcal{M}$ be a nonstandard model of $I\Delta_0$ and $a \in \mathcal{M}$, and consider the initial segment

$$BP[a] = \{x \in \mathcal{M} : x < a^n \text{ for some } n \in \mathbb{N}\}.$$

As we saw, $BP[a]$ is also a model of $I\Delta_0$. Moreover, if $\mathcal{M} \models \Delta_0\text{-PHP}$ then also $BP[a] \models \Delta_0\text{-PHP}$. But clearly, in $BP[a]$, exponentiation is not a total function. The following is still an open problem.

QUESTION 6.2 (Macintyre). *Does $I\Delta_0 \vdash \Delta_0\text{-PHP}$?*

Using a forcing argument, Ajtai showed in [1] that if we add a new functional symbol $f$ to the language then $I\Delta_0(f) \nvdash \Delta_0(f)\text{-PHP}$. So a proof of $\Delta_0\text{-PHP}$ in $I\Delta_0$ cannot come from purely logical arguments.

NOTE 6.3. We recall that DOR denotes the theory of discretely ordered rings. Then

$$\text{DOR} + \Delta_0\text{-PHP} \vdash I\Delta_0.$$

Let $\mathcal{M} \models \text{DOR} + \Delta_0\text{-PHP}$. Suppose that $\mathcal{M} \models \phi(0) \wedge \forall x(\phi(x) \rightarrow \phi(x + 1))$, where $\phi \in \Delta_0$ and $\mathcal{M} \models \neg\phi(a)$ for some $a \in \mathcal{M}$. Consider the following function $f : a + 1 \rightarrow \mathcal{M}$,

$$f(b) = \begin{cases} b & \text{if } \mathcal{M} \models \phi(b) \\ b - 1 & \text{if } \mathcal{M} \models \neg\phi(b). \end{cases}$$

Clearly, $f$ is $\Delta_0$-definable. It is also an injective function from $a + 1$ into $a$. We have then a contradiction with $\Delta_0\text{-PHP}$. This, together with Woods' result, implies that DOR $+ \Delta_0\text{-PHP}$ proves the cofinality of primes.

The following more general result holds.

THEOREM 6.4.  $I\Delta_0 + \Sigma_n\text{-PHP} \nvdash I\Sigma_n$ for $n > 0$.

PROOF. The same argument as before shows that $\Sigma_n$-PHP implies $\Sigma_n$-induction. For the converse, just recall that exponentiation is provably total in $I\Sigma_1$, and so functions can be coded. For more detail see [19].    $\dashv$

Paris, Wilkie, and Woods improved Woods' result by showing that in fact a weak version of the pigeonhole principle is enough to show that there are infinitely many primes. The principle is the following:

for all $x$ there is no 1-1 $\Delta_0$-function $f$ such that $f : 2x \longrightarrow x$.   ($\Delta_0$-WPHP)

The weak pigeonhole principle is available in the theory $I\Delta_0 + \Omega_1$ where $\Omega_1$ is the axiom

$$\forall x, y \exists z \left( x^{[\log_2 y]} = z \right).$$

The system $I\Delta_0 + \Omega_1$ has been widely studied. It has emerged as an economical system where an easy coding of syntax is possible (see [50]). From the number theoretical point of view, we do not know much about $I\Delta_0$, except, as we have seen, that we can give a meaning to unique factorization. In $I\Delta_0 + \Omega_1$, it is possible to prove cofinality of primes (see [37]), Lagrange's theorem (see [6]), and classical results about residue fields (see [16]). By an argument a la Parikh, we can show that $I\Delta_0 + \Omega_1 \nvdash \exp$. Let $\mathcal{M}$ be a model of $I\Delta_0 + \Omega_1$, $a \in \mathcal{M} - \mathbb{N}$, and consider

$$BL[a] = \left\{ x \in \mathcal{M} : x < a^{(\log a)^n} \text{ for some } n \in \mathbb{N} \right\}.$$

Clearly, $BL[a] \models I\Delta_0 + \Omega_1$, but $BL[a] \nvDash \exp$. So among the extensions of $I\Delta_0$, the following strict relations hold:

$$I\Delta_0 \subset I\Delta_0 + \Omega_1 \subset I\Delta_0 + \exp.$$

Notice that there is no clear relation between the full pigeonhole principle $\Delta_0$-PHP and $\Omega_1$.

Let # be a new function symbol where $\#(x, y) = x^{[\log_2 y]}$. We expand the language of arithmetic by #, and we denote by $\Delta_0^\#$ the class of bounded formulas in the expanded language where the bounds now are terms containing the symbol #. If we allow induction on $\Delta_0^\#$-formulas, we get the theory $I\Delta_0^\#$. It is not difficult to show that

$$I\Delta_0^\# \vdash \forall x > 0 \forall y \forall t \forall z (\#(x, y) = z \leftrightarrow (t = [\log_2 y] \wedge E_0(x, t, z))).$$

It follows that in any model $\mathcal{M}$ of $I\Delta_0^\#$ the function $f(x) = x^{[\log_2 x]}$ is total. Conversely, the function defined by

$$\theta(x, y, z) \overset{\text{def}}{=} (y = 0 \wedge z = 1) \vee \exists t < y(t = [\log_2 y] \wedge E_0(x, t, z))$$

is provably total in $I\Delta_0 + \Omega_1$ and satisfies some basic natural algebraic properties of #. We will not make any distinction whether we work in $I\Delta_0^\#$ or in $I\Delta_0 + \Omega_1$.

REMARK 6.5. (i) Obviously, $I\Delta_0^\#$ does not imply that exponentiation is a total function, but in any model of $I\Delta_0^\#$, if $2^n$ is defined, then also $2^{n^k}$ is defined for all $k \in \mathbb{N}$. Notice also that the domain of definition of exponentiation coincides in this case with the domain of definition of factorial because of the inequalities $n! \leq n^n \leq 2^{n^2}$ (these inequalities are provable in $I\Delta_0$; see next section).

(ii) # is a polynomial time computable function.

(iii) Let $x$ be an element of $\mathcal{M}$, a model of $I\Delta_0 + \Omega_1$. The domain of definition of exponentiation in basis $x$ is an initial segment closed under $+$ and $\cdot$, so by Lemma 3.1 it is a model of $I\Delta_0$. While $I\Delta_0$ can capture only the linear increase of the lengths, $I\Delta_0 + \Omega_1$ captures the polynomial increase of the lengths.

§7. **Chebyshev's theorem.** Although it is not known whether $I\Delta_0$ can prove the cofinality of primes, we have a local version of the classical result of Chebyshev on the rate of growth of the function product of primes.

THEOREM 7.1. $I\Delta_0 \vdash$

(i) $\forall x \forall y (2^x = y \rightarrow \exists z < y^2 \prod_{p \leq x} p = z)$;

(ii) $\forall x \forall y (\prod_{p \leq x} p = y \rightarrow \exists z < y^4 2^x = z)$.

We do not go through the details of the proof which can be found in [11]. We note that for the proof of Chebyshev's classical result it has been necessary to define in a $\Delta_0$-way functions such as factorial and binomial coefficient, and to prove the basic recursion laws of these functions in $I\Delta_0$. In order to define factorial, the following result due to Paris, Wilkie, and Woods has been used.

THEOREM 7.2. *Let $a, b, d \in \mathcal{M}$, $d \leq (\log a)^k$ for some $k \in \mathbb{N}$ and $F : d \rightarrow b$, $\Delta_0$-definable. Then there is a $\Delta_0$-definable function $G : d \rightarrow \mathcal{M}$ (uniformly) such that $G(0) = F(0)$ and for all $i < d$, $G(i+1) = G(i) + F(i+1)$, i.e., the sum of the function $F$ exists.*

The uniformity in Theorem 7.2 is connected with the possible presence of parameters in the formula defining $F$. In this case, the parameters are carried into the $\Delta_0$-definition of $G$. In the proof, the expansion in base 2 of the $F(i)$s is used.

In $\mathbb{N}$ factorial has the following expression

$$x! = \prod_{p \leq x} p^{\sum [x/p^i]}$$

where $\sum [x/p^i]$ is the number of multiples of powers of prime $p$ which are less than or equal to $x$. We now formalize this in $I\Delta_0$. Fix a prime $p \leq x$, and consider the $\Delta_0$-function $F_{x,p} : [\log_p x] \rightarrow x$ defined as

$$F_{x,p}(i) = \left[ \frac{x}{p^i} \right].$$

Clearly, $F_{x,p}(i) = 0$ if $i > [\log_p x]$. Function $F_{x,p}$ satisfies the hypothesis of Theorem 7.2, and so there is a $\Delta_0$-function $G_{x,p}$ which defines the $p$-adic valuation of $x!$ as the sum of the sequence

$$\left[\frac{x}{p^0}\right], \left[\frac{x}{p}\right], \ldots, \left[\frac{x}{p^k}\right],$$

for $k = [\log_p x]$. The recursion laws of factorial are provable using $\Delta_0$-induction.

THEOREM 7.3. $I\Delta_0 \vdash$

(i) $\forall x, y(x! = y \longrightarrow \forall z \le x(z \mid y))$,
(ii) $\forall x, y(x! = y \longrightarrow (x+1)! = y(x+1))$,
(iii) $\forall x, y((x+1)! = y \longrightarrow \exists z < y(x! = z \wedge z(x+1) = y))$,
(iv) $\forall x, y(x \ge 4 \wedge x! = y \longrightarrow \exists z < y2^x = z)$.

Notice that (iv) implies that factorial is not a provably total function in $I\Delta_0$. In any model of $I\Delta_0$, the domain of definition of factorial is an initial segment closed under successor and is a subset of the domain of definition of exponentiation. In general the inclusion is strict: consider a model $\mathcal{M}$ of $Th(\mathbb{N})$ and let $n \in \mathcal{M} - \mathbb{N}$. The structure $BP[2^n]$ is a model of $I\Delta_0$ in which $n!$ is not defined. If so then $n! < (2^n)^k$ for some $k \in \mathbb{N}$. Since $\mathcal{M} \models Th(\mathbb{N})$ Stirling's formula, i.e.,

$$\sqrt{2\pi n}\, n^{\frac{n+1}{2}} e^{\frac{-n-1}{2}} \le n! \le \sqrt{2\pi n}\, n^{\frac{n+1}{2}} e^{\frac{-n+1}{2}},$$

(or rather a suitable formalization of it to handle $e$) is true in $\mathcal{M}$. Hence $n^{n+1} \le (2^n)^k$ if and only if

$$n^{n+1} \le 2^{2nk} 3^{n+1} < 3^{2nk+n+1} < 3^{2nk+n+2k+1}$$
$$= 3^{n(2k+1)+2k+1} = \left(3^{(2k+1)}\right)^{n+1}.$$

By properties of the exponential function, we get $n < 3^{2k+1} \in \mathbb{N}$, which is a contradiction.

In order to guarantee the existence of $n!$, it is, however, enough to assume that $2^{n^2}$ exists. In a model of $I\Delta_0$ where exponentiation is total, factorial also is a total function.

Now that we have an unambiguous definition of factorial, we can also give a meaning to the binomial coefficient in any nonstandard model of $I\Delta_0$ in such a way as to reproduce the binomial theorem. In $\mathbb{N}$,

$$\binom{n}{k} = \frac{n!}{k!(n-k)!}$$

for any $n, k$ with $k \le n$. So we define the binomial coefficient $\binom{n}{k}$ as the lcm of the set $\{p^j : p \le n, p \text{ is a prime}, j = v_p(n!) - v_p(k!) - v_p((n-k)!)\}$. Obviously, if such a $y$ exists, then it is unique, and so we will use the standard notation $\binom{n}{k}$.

LEMMA 7.4. $I\Delta_0 \vdash$

(i) $\binom{n}{k} = y \to \exists z, w(z = \binom{n+1}{k} \wedge w = \binom{n}{k-1})$,
(ii) $(n - k + 1)\binom{n+1}{k} = \binom{n}{k}(n + 1)$,
(iii) $k\binom{n}{k} = \binom{n}{k-1}(n - k + 1)$,
(iv) $\binom{n+1}{k} = \binom{n}{k} + \binom{n}{k-1}$.

Using Lemma 7.4 we can prove the binomial theorem in every model $\mathcal{M}$ of $I\Delta_0$ for all elements belonging to $B_2$, the domain of definition of base 2 exponentiation.

THEOREM 7.5. Let $n \in B_2$ and $m = 2^n$. Then $\mathcal{M} \models \forall k \le n \exists x < m(\binom{n}{k} = x \wedge m = \sum_{k=0}^{n}\binom{n}{k})$.

From Theorem 7.5 and Lemma 7.4, the following theorem can be proved.

THEOREM 7.6. Suppose that $\binom{2n}{n} = x$ for some $x \in \mathcal{M}$. Then there exists $m$ such that $m \le x$ and $m = 2^n$.

REMARK 7.7. Notice that $2^n$ is defined if and only if $\binom{2n}{n}$ is defined, and the following inequalities hold: $2^n \le \binom{2n}{n} \le 2^{2n}$.

Now we have all the ingredients in order to prove Chebyshev's result, i.e., also in $I\Delta_0$, the two functions $2^x$ and $\prod_{p \le x} p$ have the same rate of growth. Hence, they have the same domain of definition. Notice that there are models of $I\Delta_0$ where the primes run through to the end of the model, and the function $\prod_{p \le x} p$ is not total (just consider a model of $\Omega_1$ but not of exp).

From these results, it follows that in every model of $I\Delta_0$ a local version of Bertrand's postulate holds.

COROLLARY 7.8. For every $x \in B_2$ there exists a prime $p$ such that $x < p \le 2x$.

The same ideas used for defining factorial have been used by Berarducci and D'Aquino in [5] to prove that if $F$ is a $\Delta_0$-definable function, then the product function,

$$\begin{cases} G(0) = F(0) \\ G(i + 1) = G(i)F(i + 1) \quad \text{for all } i < n, \end{cases}$$

is also $\Delta_0$-definable. Notice that except for the $\Delta_0$-definability of $F$, there is no further hypothesis on the function. We like to stress the contrast with the sum function of a $\Delta_0$-definable function, i.e., the function defined as

$$\begin{cases} S(0) = F(0) \\ S(i + 1) = S(i) + F(i + 1) \quad \text{for all } i < n. \end{cases}$$

In Theorem 7.2, under the further hypothesis that the function values are bounded and $n$ is logarithmic it is proved that $S$ is in turn $\Delta_0$-definable. We briefly discuss the definition of the product function. Clearly,

$$y = \prod_{i \leq n} F(i) \text{ if and only if } y = \prod_{p \leq y} p^{\sum_{i \leq n} v_p(F(i))}. \tag{8}$$

If $F(i) = 1$ for *too many* $i < n$'s then $n$ may not be logarithmic and we cannot apply Theorem 7.2 in order to define the $p$-adic valuation of $\prod_{i \leq n} F(i)$. We give the following equivalent definition of $p$-adic valuation of $y = \prod_{i \leq n} F(i)$.

$$v_p(y) = \sum_{h \leq \log y} h |\{i \leq n : v_p(F(i)) = h\}| \tag{9}$$

where $|A|$ denotes the cardinality of a set $A$. In other words, we count the number of $i$s less than $n$ for which $F(i) > 1$. In this way we reduce the computation of $G(n) = \prod_{i \leq n} F(i)$ to the case when $F$ is always $> 1$ and therefore $G$ grows exponentially in $n$. This approach needs to be handled with some care since in $I\Delta_0$ we cannot *count*; there is not a good notion of cardinality. This is again related to the problem of the nontotality of exponentiation: in order to formalize the notion of cardinality we need to use functions and we may not be able to code them. The following is still an open problem.

QUESTION 7.9 (Paris and Wilkie). *Given $A \subseteq \mathbb{N}$ and $A \in \Delta_0^{\mathbb{N}}$, is the function $H(n) = |A \cap n|$ in $\Delta_0^{\mathbb{N}}$?*

Paris and Wilkie in [36] give a positive answer in the case $|A \cap n| \leq (\log n)^k$, $k \in \mathbb{N}$, and they formalize it also in $I\Delta_0$. The problem of defining $|\{i \leq n : v_p(F(i)) = h\}|$ is then solved in [5] by an improved version of Paris and Wilkie's result. In a model $\mathcal{M}$ of $I\Delta_0$, it may even happen that a set $A$ has two different cardinalities: if there are bijections from $A$ into $m$ and $n$ then we may not be able to prove that $n = m$, as far as we know. The pigeon-hole principle avoids this. If the model satisfies the pigeonhole principle, and a set has a cardinality, then this is unique. In $I\Delta_0$, the following logarithmic version of the pigeonhole principle is provable:

$$\forall x \neg \exists \ 1\text{-}1 \ \Delta_0\text{-function } f$$

$$\text{such that } f : (\log x)^k + 1 \longrightarrow (\log x)^k \quad (\log -\Delta_0\text{-PHP})$$

for all $k \in \mathbb{N}$. Paris and Wilkie use $\log -\Delta_0$-PHP in the formalization in $I\Delta_0$ of their definition of $|A \cap n| \leq (\log n)^k$, $k \in \mathbb{N}$.

§8. **Pell equations.** Pell equations played a crucial role in the proof of the fundamental theorem due to Matijasevič, Robinson, Davis, and Putnam, which asserts that every r.e. set is existentially definable. We will use the notation MRDP-theorem to denote this result. This theorem led to the negative solution of Hilbert's tenth Problem. One of the main open problems in weak fragments of arithmetic is the following.

QUESTION 8.1. *Does $I\Delta_0 \vdash$ the MRDP-theorem?*

We can reformulate this problem as follows: for every $\Sigma_1$-formula $\theta(\bar{x})$ are there polynomials $p(\bar{x}, \bar{y})$ and $q(\bar{x}, \bar{y})$ such that

$$I\Delta_0 \vdash \forall \bar{x}(\theta(\bar{x}) \leftrightarrow \exists \bar{y}(p(\bar{x}, \bar{y}) = q(\bar{x}, \bar{y}))?$$

The interest in studying this problem arises from the connections with complexity theory. As Wilkie observed, a positive solution to this problem would solve in a positive way the well-known open problem of whether NP = co-NP. This is a consequence of the following two facts:

(i) the set $H = \{(a, b, c) \in \mathbb{N}^3 : \exists x < c \exists y < c(ax^2 + by = c)\}$ is NP-complete, as shown by Adleman and Manders (see [27]);

(ii) if $I\Delta_0 \vdash$ MRDP-theorem, then $\Delta_0^\mathbb{N} = E_1^\mathbb{N}$, and this implies that $H \in \text{NP} \cap \text{co-NP}$.

By $\Delta_0^\mathbb{N}$ and $E_1^\mathbb{N}$ we denote the subsets of $\mathbb{N}$ which are defined by a $\Delta_0$-formula and an $E_1$-formula (only existentially bounded quantifiers appear in the formula), respectively.

From (i) and (ii) it follows that NP = co-NP.

It is also unknown whether the theory $I\Delta_0 + \Omega_1$ proves the MRDP-theorem. A positive answer would also show NP = co-NP. On the other hand, Gaifman and Dimitracopoulos proved that $I\Delta_0 + \exp \vdash$ MRDP-theorem (see [18]), but this does not have any consequence in complexity theory.

We now study the theory of Pell equations over $I\Delta_0$ and $I\Delta_0 + \Omega_1$.

A Pell equation is of the form $x^2 - dy^2 = 1$, where $d$ is not a square (denoted by $d \neq \square$). The theory of the solutions of a Pell equation is completely understood in $\mathbb{N}$:

(i) there is always a nontrivial solution, i.e., a solution $(x, y) \neq (1, 0)$;

(ii) the set of solutions has a cyclic group structure, and a generator is called a fundamental solution. If $(x_1, y_1)$ is the minimal positive nontrivial solution, then all other solutions are obtained as $(x_1 + \sqrt{d} y_1)^n$, for $n \in \mathbb{Z}$.

Define

(P) Any Pell equation has a nontrivial solution.

First we consider the strength of axiom (P), and we compare it with exp over $I\Delta_0$. The proof of (P) usually proceeds by showing the existence of nontrivial units in the quadratic extension $\mathbb{Q}(\sqrt{d})$.

First of all we need to extend some notions regarding quadratic extensions to a model $\mathcal{M}$ of $I\Delta_0$. By considering pairs of elements of $\mathcal{M}$, we can give a meaning to the notion of quadratic extension also in this setting. In a natural way, we can extend the operations of $+$ and $\cdot$ to $\mathcal{M}[\sqrt{d}]$. It is also convenient to define an order on $\mathcal{M}[\sqrt{d}]$, by

$$p + \sqrt{d}q < r + \sqrt{d}s \quad \text{iff} \quad (p - r)^2 < d(s - p)^2.$$

Next, we extend the notion of integer part of $x + \sqrt{d}\, y \in M[\sqrt{d}]$:

$[x + \sqrt{d}\, y] = z$ iff $z$ is the greatest element of $M$ satisfying $z \leq x + \sqrt{d}\, y$ (or equivalently $(z - x)^2 \leq dy^2$).

This definition makes sense because of Lemma 3.4, where $A = \{w : (w - x)^2 \leq dy^2\}$ is clearly $\Delta_0$-definable and bounded by $x + dy$.

It makes sense now to talk about the fractional part of an element of $M[\sqrt{d}]$. We define $\{x + \sqrt{d}\, y\} = x + \sqrt{d}\, y - [x + \sqrt{d}\, y]$.

In a natural way, we can define the norm of an element $x + \sqrt{d}\, y$ of $M[\sqrt{d}]$ by $N(x + \sqrt{d}\, y) = x^2 - dy^2$. Next we formalize the notion of $n$th power of an element $x + y\sqrt{d}$ of $M[\sqrt{d}]$. We define the relation $(x + y\sqrt{d})^n = u + v\sqrt{d}$ as follows

$$u = \sum_{\substack{i \leq n \\ i \equiv 0(2)}} \binom{n}{i} x^{n-i} y^i d^{i/2} \quad \text{and} \quad v = \sum_{\substack{i \leq n \\ i \equiv 1(2)}} \binom{n}{i} x^{n-i} y^i d^{(i-1)/2}. \quad (10)$$

As we saw, exponentiation and binomial coefficient are both $\Delta_0$-definable. Moreover, $n$ is logarithmic, and so by Theorem 7.2, it follows that the relations in (10) are $\Delta_0$. Exponentiation in $M[\sqrt{d}]$ is also a partial function, but the usual properties of exponentiation can be proved in $I\Delta_0$ (for the details see [13]). It is straightforward to extend the notions above to the ring $R_M$ associated to $M$. We now define the notion of negative power in a restricted context.

DEFINITION 8.2. Let $x, y, n \in M$, and $x^2 - dy^2 = 1$. Define

$$(x + y\sqrt{d})^{-1} = x + (-y)\sqrt{d}, \text{ and } (x + y\sqrt{d})^{-n} = ((x + y\sqrt{d})^n)^{-1}.$$

Note that $(x + \sqrt{d}\, y)^{-1} \in R_M[\sqrt{d}]$ and is the ring inverse.

We now examine the proof of $(P)$ based on Dirichlet's Theorem on diophantine approximation which uses a pigeonhole argument. In fact we only need a weak version of the pigeonhole principle. This will enable us to prove a version of Dirichlet's theorem in $I\Delta_0 + \Omega_1$.

THEOREM 8.3. Let $M \models I\Delta_0 + \Omega_1$, $d \in M$, $d$ not a square, $Q > 1$. Then there are $p, q \in M$ such that $|p - \sqrt{d}\, q| < 1/Q$, and $q < 2Q$.

PROOF. Axiom $\Omega_1$ implies $\Delta_0$-WPHP, i.e., $M \models \neg\exists$ 1-1 $f : [0, 2a] \to [0, a]$ for all $a \in M$. Let $h : [0, 2Q] \to [0, 1]$ be defined as follows: $h(n) = \{n\sqrt{d}\}$. $h$ is $\Delta_0$-definable, and by an easy calculation it is injective. Let $g$ be a partition of the interval $[0, 1]$ into $Q$ equal parts, i.e., $g : [0, Q] \to [0, 1]$, defined as $g(i) = i/Q$. Clearly $g$ is $\Delta_0$-definable.

By $\Delta_0$-WPHP, we cannot fit the elements of $[0, 2Q]$ into the $Q$ boxes in which $[0, 1]$ has been divided by $g$. So there must exist $n, k \leq 2Q$ with $k < n$ such that $|\{n\sqrt{d}\} - \{k\sqrt{d}\}| < 1/Q$. But $|\{n\sqrt{d}\} - \{k\sqrt{d}\}| = |n\sqrt{d} - [n\sqrt{d}] - k\sqrt{d} + [k\sqrt{d}]|$, so if $p = [n\sqrt{d}] - [k\sqrt{d}]$ and $q = n - k$, it follows that $|p - q\sqrt{d}| < 1/Q$ and $q < 2Q$.                                                                                  ⊣

Clearly, $p, q$ can be chosen coprime. Notice that we can also bound the size of $p$, since $|p| = |p - \sqrt{d}q + \sqrt{d}q| \leq |p - q\sqrt{d}| + q\sqrt{d} < 1/Q + 2Q\sqrt{d} \leq 3Q\sqrt{d}$.

So what the theorem shows is

$$I\Delta_0 + \Omega_1 \vdash \forall d\forall Q(d \neq \square \wedge Q > 1$$
$$\rightarrow \exists p \leq 3Q\sqrt{d}\exists q \leq 2Q|p - q\sqrt{d}| < 1/Q).$$

With simple algebraic calculations, one can show that the element $\alpha = p - q\sqrt{d}$ of $\mathcal{M}[\sqrt{d}]$ has norm bounded independently from the choice of $Q$:

$$|N(p - q\sqrt{d})| = |(p - q\sqrt{d})(p + q\sqrt{d})| \leq |p - q\sqrt{d}|(|p - q\sqrt{d}| + 2q\sqrt{d}|)$$
$$< (1/Q)(1/Q + 4Q\sqrt{d}) \leq (1/Q)5Q\sqrt{d} = 5\sqrt{d}.$$

The proof of $(P)$ in $\mathbb{N}$ proceeds now with an iteration of Dirichlet's theorem in order to get infinitely many different pairs $(p, q)$ such that $N(p - q\sqrt{d}) < 5\sqrt{d}$. To be more precise, let $(p, q)$ be the pair of integers associated to a starting $Q$. Choosing now $Q_1 > \frac{1}{|p - q\sqrt{d}|}$, we get $(p_1, q_1)$ satisfying $|p_1 - q_1\sqrt{d}| < 1/Q_1$ and moreover, $p/q \neq p_1/q_1$. The last inequality is true because of the irrationality of $\sqrt{d}$.

Since the norms of all elements $p + \sqrt{d}q$ are bounded by $5\sqrt{d}$, by a pigeonhole argument, there are infinitely many pairs $(p, q)$ such that $N(p + q\sqrt{d}) = N$, for some $N < 5\sqrt{d}$. Among these, there will be at least two pairs $(p, q)$ and $(p_1, q_1)$ such that $p \equiv p_1 \pmod{N}$ and $q \equiv q_1 \pmod{N}$. A linear combination of $p, p_1, q, q_1, N$ gives a nontrivial solution of $X^2 - dY^2 = 1$ (see [3]).

It is not in fact necessary to have infinitely many pairs $(p, q)$ constructed as above. With a right use of the pigeonhole principle, we could obtain the same result using only finitely many pairs. It is clear that in this procedure, a recursive argument is hidden which we need to code in $I\Delta_0 + \Omega_1$ if we want to reproduce the proof of $(P)$ in such a fragment. As already remarked, $\Omega_1$ does not offer any guarantee that this coding is possible. On the other hand, if exponentiation is total, then the coding is possible, and we can reproduce the proof of $(P)$, as Dimitracopoulos showed in [17]. The result we will prove later confirms that $\Omega_1$ is not enough to prove $(P)$.

Given a Pell equation

$$X^2 - dY^2 = 1 \qquad (d \neq \square), \tag{11}$$

via simple algebraic calculations, one can easily prove that the product of two solutions is still a solution of (11), i.e., if $(u, v)$ and $(w, t)$ are solutions of (11), then the product of them (as an element of $\mathcal{M}[\sqrt{d}]$) is also a solution; i.e., $(u + \sqrt{d}v)(w + \sqrt{d}t) = uw + dvt + \sqrt{d}(ut + vw)$ and $(uw + dvt)^2 - d(ut + vw)^2 = 1$. Using the $\Delta_0$-definition of $n$th power of an element of $\mathcal{M}[\sqrt{d}]$, we can extend the notion of $n$th solution of (11) also to models of $I\Delta_0$ in the following sense. Assume that (11) has a nontrivial solution $(x_1, y_1)$ (without loss of generality we can assume that $(x_1, y_1)$ is minimal). Then the pair $(x_n, y_n)$ obtained as $x_n + \sqrt{d}\,y_n = (x_1 + \sqrt{d}\,y_1)^n$ is the $n$th solution. As already remarked, there are restrictions on the existence of the $n$th solution. The following theorem states that for those solutions which exist, the same group structure as in the standard case holds. Moreover, any solution of (11) is obtained as a power of the fundamental solution. Let $\mathcal{M} \models I\Delta_0$.

THEOREM 8.4.

(i) $\mathcal{M} \models \forall n \forall u \forall v (u + v\sqrt{d} = (x_1 + y_1\sqrt{d})^n \rightarrow u^2 - dv^2 = 1)$.

(ii) $\mathcal{M} \models \forall u \forall v (u^2 - dv^2 = 1 \rightarrow \exists n \leq u (x_1 + y_1\sqrt{d})^n = u + v\sqrt{d})$.

(iii) *There is no solution $(u, v)$ of $(\star)$ satisfying $(x_1 + y_1\sqrt{d})^n < u + v\sqrt{d} < (x_1 + y_1\sqrt{d})^{n+1}$.*

REMARK 8.5. For some Pell equations, there is no problem in finding a nontrivial solution. For example, the equation $X^2 - (a^2 - 1)Y^2 = 1$, where $a \geq 2$, is satisfied by the pair $(a, 1)$. Let $(x_n(a), y_n(a))$ denote the $n$th solution of $X^2 - (a^2 - 1)Y^2 = 1$. From now on, we will omit $a$ if it is clear from the context of the equation we are working with. From the definition of $n$th solution it is clear that $x_n(a)$ and $y_n(a)$ are defined iff $a^n$ is defined.

Let $a \in \mathcal{M} \models I\Delta_0$. The following properties of the solutions of the Pell equation $X^2 - (a^2 - 1)Y^2 = 1$ hold in $\mathcal{M}$.

LEMMA 8.6.

(1) $\forall n \forall x \forall y (x = x_n \wedge y = y_n \rightarrow \sqrt{a^2 - 1}\,y \leq x \leq ay)$.

(2) $\forall n \forall y (y = y_{n+1} \rightarrow \exists z \leq y^2 \exists w \leq y$
$\qquad\qquad\qquad (z = (2a)^n \wedge w = (2a - 1)^n \wedge w \leq y \leq z))$.

(3) $\forall n \forall x (x = x_n \rightarrow x \equiv 1 \,(\mathrm{mod}\, a - 1))$.

(4) $\forall n \forall y (y = y_n \rightarrow y \equiv n \,(\mathrm{mod}\, a - 1))$.

(5) $\forall n \forall y (y = y_n \rightarrow y \geq n)$.

(6) $\forall n \forall m \forall y \forall z (y = y_n \wedge z = y_m \wedge y \mid z \rightarrow n \mid m)$.

(7) $\forall n \forall m \forall y \forall z (y = y_n \wedge z = y_m \wedge y^2 \mid z \rightarrow y \mid m)$.

These properties were first proved in $\mathbb{N}$ by Matijasevič in order to give a diophantine definition of exponentiation. For the proof in $I\Delta_0$ see [13]. Properties (6) and (7) hold for a general Pell equation while the previous ones are specific for Pell equations of the form $X^2 - (a^2 - 1)Y^2 = 1$.

We can now prove the following result.

THEOREM 8.7. $I\Delta_0 + P \vdash \exp$.

PROOF. Let $\mathcal{M} \models I\Delta_0 + P$. We want to show that exponentiation is a total function on $\mathcal{M}$, i.e., for any $a > 0, c \in \mathcal{M}$ there is an element $b$ in $\mathcal{M}$ satisfying $b = a^c$. We recall that for any fixed base, the logarithmic function is a total function in $\mathcal{M}$. So there is $n \in \mathcal{M}$ such that $a^n \leq c < a^{n+1}$. Consider now the Pell equation

$$X^2 - (a^2 - 1)Y^2 = 1. \tag{12}$$

The existence of $a^{n+1}$ implies that there is $y \in \mathcal{M}$ satisfying $y = y_{n+1}(a)$; let $x = x_{n+1}(a)$. Now consider the following Pell equation:

$$X^2 - (a^2 - 1)(2x^2y^2)^2 Y^2 = 1. \tag{13}$$

From $\mathcal{M} \models P$, it follows that (13) has a nontrivial solution in $\mathcal{M}$, i.e., there are $u, v \in \mathcal{M}$ satisfying $u^2 - (a^2 - 1)(2x^2y^2)^2v^2 = 1$, and $v > 0$. Let $w = u$ and $z = 2x^2y^2v$. We can show that $(w, z)$ is a solution of (12), i.e., $w^2 - (a^2-1)z^2 = u^2 - (a^2-1)(2x^2y^2)^2v^2 = 1$, where the last equality is true since $(u, v)$ is a solution of (13). From the cyclic group structure of the set of solutions it follows that there is $k \in \mathcal{M}$ such that $(u, z) = (x_k(a), y_k(a))$. Notice that the solutions $y_{n+1}(a)$ and $y_k(a)$ satisfy $y_{n+1}^2(a) \mid y_k(a)$; hence by (7) of Lemma 8.6, $y_{n+1}(a) \mid k$. The existence of the $k$th solution of (12) implies that $a^k$ is defined in $\mathcal{M}$, and from $y_{n+1}(a) \leq k$ it follows that $a^{y_{n+1}(a)}$ must also be defined. But $y_{n+1}(a) \geq a^{n+1}$, which implies $a^{a^{n+1}}$ is defined. Since $c < a^{n+1}$ we can finally deduce that $a^c$ is defined.                    $\dashv$

REMARK 8.8. Combining Dimitracopoulos's result that $I\Delta_0 + \exp \vdash P$ and Theorem 8.7, we get that the axioms exp and $(P)$ are equivalent over $I\Delta_0$. This does not support the idea of Jones and Matijasevič in [21] that axiom $(P)$ is weaker than exp.

Theorem 8.7 confirms that $\Omega_1$ is not enough to prove the existence of a nontrivial solution for any Pell equation. If we add $\Delta_0$-PHP to $I\Delta_0$, we get a theory which is not strong enough to prove axiom $(P)$. Otherwise we would get that $\Delta_0$-PHP implies exponentiation over $I\Delta_0$, which is false.

Theorem 8.7 can be improved in the following sense: Kaye in [22] studied the theory $IE_1$ where induction is applied only to formulas with bounded existential quantifiers. It is still an open problem if the theories $I\Delta_0$ and $IE_1$ coincide. This is related to the problem of the collapse of the $\Delta_0$-hierarchy. Kaye considered the following axiom,

$$\forall a \geq 2 \forall b \leq a - 2 \exists u \exists v (u^2 + v^2 - 2auv - 1 = 0 \wedge u \leq v$$
$$\wedge u \equiv b \pmod{a-1} \wedge y \equiv b + 1 \pmod{a-1}), \quad (E)$$

and proved that if we add $(E)$ to $IE_1$, then we obtain a system equivalent to $I\Delta_0 + \exp$. Axiom $(E)$ guarantees the existence of a solution of the Pell equation $X^2 - (a^2 - 1)Y^2 = 1$ in each class modulo $a - 1$; the one corresponding to class $a - 2$ will be *big*, in the sense that it is at least of order $a^{a-2}$. So axiom $(E)$ says, roughly speaking, that $a^a$ has to be defined.

On the one hand, it is quite remarkable that the $E_1$-induction can be made to capture the power of $\Delta_0$-induction. On the other hand, from what we observed it is not so surprising that axiom $(E)$ implies exponentiation is total.

Using Kaye's result we strength Theorem 8.7 as follows.

THEOREM 8.9. $IE_1 + P \vdash I\Delta_0 + \exp$.

For the proof see [13].

We now examine the solvability of Pell equations over residue rings of models of $I\Delta_0$. Let $\mathcal{M}$ be a model of $I\Delta_0$ and $d, p \in \mathcal{M}$ where $p > 2$ is a prime and $(d, p) = 1$. From Hilbert's Theorem 90 it follows that the equation $X^2 - dY^2 = 1$ has a nontrivial solution in the quotient field $\mathcal{M}/\langle p \rangle$ (here by a nontrivial solution, we mean a solution $x \not\equiv \pm 1 \pmod{p}$).

Via just algebraic calculation, we can show that any solution modulo $p^n$ can be lifted up to a solution modulo $p^{n+1}$ (Hensel's Lemma).

Suppose now that $m = p_1^{k_1}, \ldots, p_s^{k_s}$, $(d, m) = 1$ and $d \not\equiv \square \pmod{p_i}$ for $i = 1, \ldots, s$. Let $(u_i, v_i)$ be a nontrivial solution of $X^2 - dY^2 = 1$ modulo $p_i^{k_i}$ for $i = 1, \ldots, s$. Using the Chinese Remainder Theorem we can solve the two systems of congruences $x \equiv u_i \pmod{p_i^{k_i}}$ and $y \equiv v_i \pmod{p_i^{k_i}}$, which give a solution of $X^2 - dY^2 = 1$ in the residue ring $\mathcal{M}/\langle m \rangle$.

A local-global principle for nontrivial solvability of Pell equations does not hold in $I\Delta_0$: no exponentiation is needed in order to solve nontrivially in a residue ring, while we saw that exponentiation is equivalent to solvability in a nontrivial way over the whole model. For details see [14].

J. Robinson was the first to link the theory of Pell equations to definability problems. Her ideas were developed by Matijasevič and their work led to an existential definition of the exponential function, which is one of the main steps in the proof that every r.e. set is existentially definable. Their definition is given in terms of solutions of Pell equations. On the other hand, we saw that there is a $\Delta_0$-definition of exponentiation. We are now interested in a common refinement of the two definitions from the point of view of complexity of defining formula, i.e., we will try to define the relation $a^n = m$ using only existentially bounded quantifiers.

An attempt to modify Paris's definition of exponentiation is not sensible, since that definition involves the notion of primality, and it is not known if we can express such a notion in an $E_1$-way.

We recall the existential definition of Robinson and Matijasevič, as presented in Manin's book [28]:

$$a^n = m \text{ if and only if } m = [y_{n+1}(Na)/y_{n+1}(N)]$$

for $a > 1$, $N > 4nm$, where $y_{n+1}(Na)$ is the $(n + 1)$st solution of the equation $X^2 - (N^2a^2 - 1)Y^2 = 1$ and $y_{n+1}(N)$ is the $(n + 1)$st solution of the equation $X^2 - (N^2 - 1)Y^2 = 1$.

Obviously, such a definition presupposes an existential definition of the $n$th solution of the Pell equation $X^2 - (a^2 - 1)Y^2 = 1$. The relation $y = y_n(a)$ has been uniformly defined for all $a, n, m \in \mathbb{N}$ by Matijasevič using some number theoretic properties of the solutions of a Pell equation which involve functions of double exponential growth. He uses Pell equations $X^2 - AY^2 = 1$ for $A$ of the order $a^{a^n}$. There is no hope of bounding the quantifiers over elements of double exponential size. But for our purposes, we do not need a uniform definition of the $n$th solution. The relation on $a, n, y$,

$$\exists x(x^2 - (a^2 - 1)y^2 = 1 \land y \equiv n(\bmod a - 1)), \tag{14}$$

is $E_1$-definable, but it does not give $y$ as a function of $n, a$.

The following picture, which is easily constructed using the structure (in terms of congruence modulo $n$, see Lemma 8.6) of the set of solutions, should clarify the situation:

| 0 | 1 | 2 | . | . | . | . | . | $a - 2$ |
|---|---|---|---|---|---|---|---|---|
| . | . | . | | | | | | . |
| . | . | . | | | | | | . |
| . | . | . | | | | | | . |
| $y_{-(a-1)}$ | $y_{-(a-2)}$ | $y_{-(a-3)}$ | . | . | . | . | . | $y_{-1}$ |
| $y_0$ | $y_1$ | $y_2$ | . | . | . | . | . | $y_{a-2}$ |
| $y_{a-1}$ | $y_a$ | $y_{a+1}$ | . | . | . | . | . | $y_{2a-3}$ |
| $y_{2a-2}$ | $y_{2a-1}$ | $y_{2a}$ | . | . | . | . | . | $y_{3a-4}$ |
| . | . | . | | | | | | . |
| . | . | . | | | | | | . |
| . | . | . | | | | | | . |

Relation (14) identifies infinitely many solutions for each class. Recall that the $n$th solution is roughly speaking of the order $a^n$. The idea is now to bound the size of the solution in order to pick the smallest positive solution in class $n$. In order to do this, we work in the language expanded by #, where $\#(x, y) = x^{[\log_2^y]}$ for all $y > 0$. The next claim gives a bound on the size of the solutions of the Pell equations involved in the Robinson-Matijasevič definition.

CLAIM 8.10. *Let* $m = a^n$, *and* $N = 2nm^2$. *The formula*

$$\theta(a, y, n, m) = \exists x \leq ay[x^2 - (N^2a^2 - 1)y^2 = 1$$
$$\land y \equiv n+1(\bmod Na - 1)] \land 0 < y < (\#(m, m))^6$$

*uniquely identifies the smallest positive solution in class* $n + 1$.

PROOF. Recall that $(2Na - 1)^n \leq y_{n+1}(Na) \leq (2Na)^n$, and $(2Na)^n \leq (2 \cdot 2na^{2n}a)^n \leq m \cdot m \cdot m^n \cdot m^{2n} \cdot m \leq m^3 \cdot m^{3n} \leq m^{6n} \leq (\#(m, m))^6$.

On the other hand, the next solution in class $n + 1$ is $y_{n+Na}(Na)$ (see scheme above) and $y_{n+Na}(Na) \geq (2Na - 1)^{n+Na-1} \gg m^{m^2} > (m^{[\log m]})^6 = (\#(m, m))^6$. ⊣

The same bound $(\#(m, m))^6$ also uniquely identifies the smallest positive solution in class $n + 1$ for the equation $X^2 - (N^2 - 1)Y^2 = 1$. Let $\Gamma(a, n, m)$

denote the following formula:

$$\exists y_1 < (\#(m,m))^6 \, \exists x_1 < m^5 (\#(m,m))^6 \, \exists y_2 < (\#(m,m))^6$$
$$\exists x_2 < m^4 (\#(m,m))^6 (N = 2nm^2 \wedge x_1^2 - (N^2 a^2 - 1) y_1^2 = 1 \wedge x_2^2 - (N^2 - 1) y_2^2 = 1$$
$$\wedge \, y_1 \equiv n + 1 (\mathrm{mod}\, Na - 1) \wedge y_2 \equiv n + 1 (\mathrm{mod}\, N - 1) \wedge m = [y_1/y_2]).$$

Then $a^n = m$ if and only if $\Gamma(a, n, m)$.

We can then prove in $I\Delta_0 + \Omega_1$ that $\Gamma(x, y, z)$ is equivalent to the $\Delta_0$-formula $E_0(x, y, z)$ due to Paris. We then also get that $\Gamma$ satisfies the recursion laws of exponentiation in $I\Delta_0 + \Omega_1$.

REMARK 8.11. The formula $\Gamma$ is the least complex one known so far that defines the exponential function, at the cost of having added the functional symbol #, which is a polynomial computable function.

This definition can be considered as the first step toward the proof of MRDP-theorem in $I\Delta_0 + \Omega_1$.

In the proof of Robinson and Matijasevič, there are diophantine definitions of binomial coefficient and of factorial. For these functions, it is still not known if they can be defined using only existentially bounded quantifiers. There are only partial results. We recall Robinson's definitions

$$\binom{n}{k} = m \quad \text{iff} \quad m = \mathrm{Rem}([(u + 1)^n / u^k], u) \text{ for } u > n^k, \tag{15}$$

and

$$k! = m \quad \text{iff} \quad m = [n^k / \binom{n}{k}] \text{ where } n > (2k)^{k+1}. \tag{16}$$

Clearly, functions of exponential growth are involved.

Consider first the binomial coefficient and, in particular, the case of $\binom{2n}{n}$. Let $\binom{2n}{n} = m$. Recall that in $\mathbb{N}$ the following relation holds:

$$2^n \le \binom{2n}{n} < 2^{2n}. \tag{17}$$

A local version of it can be proved in $I\Delta_0$. Relation (17) implies that for any $a$, $a^n \le \#(a, 2^n) \le \#(a, m)$. Since we work in $\mathbb{N}$, there is no problem of existence of exponentials, so we only need to set the right bounds. In J. Robinson's definition, we can choose any $u$ such that $(2n)^n < u \le 2^{n^2} \le (\#(m, m))^2$.

The next step is to bound the size of $(u + 1)^{2n}$. The following inequalities are true:

$$(u + 1)^{2n} \le (2u)^{2n} \le 2^{2n} u^{2n} < m^2 (\#(u, m))^2 \le m^2 (\#(\#(m, m)^2, m))^2$$
$$= m^2 (\#(\#(m, m), m) \cdot \#(\#(m, m), m))^2 = m^2 (\#(\#(m, m), m))^4.$$

The same bound works also for $u^n$. The notions of exponentiation, remainder, and integer part are all $E_1^\#$-definable. The above definition of $\binom{2n}{n} = m$ can be carried on in $I\Delta_0 + \Omega_1$. The relation (17) guarantees that $2^n$ is defined, and so the same procedure works. Notice that if $2^n$ is defined in $\mathcal{M} \models I\Delta_0 + \Omega_1$, then $2^{n^k}$ are defined for all $k \in \mathbb{N}$, and also $n^n$ is defined. This follows from the fact that we can define $2^{n^2} = \#(2^n, 2^n)$, $2^{n^3} = \#(\#(2^n, 2^n), 2^n), \ldots$ and $n^n = \#(n, 2^n)$.

We now examine the general case of $\binom{n}{k}$. Wlog, we can assume $k \le [n/2]$, since if $k > [n/2]$ then $\binom{n}{k} = \binom{n}{k-[n/2]}$. The following relation holds:

$$\binom{n}{i} \le \binom{n}{k} \qquad \text{for all } i \le k \le [n/2]. \tag{18}$$

We first show how to reduce the size of $u$ in (15). Consider the expression

$$\frac{(u+1)^n}{u^k} = \sum_{i=0}^{k-1} \binom{n}{i} u^{i-k} + \binom{n}{k} + \sum_{i=k+1}^{n} \binom{n}{i} u^{i-k}.$$

We want $\sum_{i \le k-1} \binom{n}{i} u^{i-k} < 1$. It is enough if $\binom{n}{i} u^{i-k} < 1/k$ for all $i \le k - 1$. By (15), it is enough that $\binom{n}{k} u^{-1} < 1/k$, i.e., $u > mk$. The problems are still unsolved in bounding the size of $u^n$. Relation (17) is not true anymore, so we cannot bound an exponential in $n$ in terms of $\binom{n}{k}$, since $\binom{n}{k}$ may be too small.

We now consider the relation $k! = m$, defined as in (16).

Even if we assume that there is an $E_1^\#$-definition of $\binom{n}{k}$, we still have problems in bounding the size of it in (16). From $2^k \le k! = m$ it follows that $k^k \le \#(k, m)$. So in (16) we can choose any $n$ such that $\#(2k, m^2) < n \le \#(m, m^2)$, and $n^k$ can be bounded by $\#(n, m^2)$. Two problems are left: the definition of $\binom{n}{k}$ and its size. Concerning the size, we know that $\binom{n}{k} < 2^{2k^{2k}}$ and we cannot express double exponential in terms of $\#$. If we expand the language $\mathcal{L}^\#$ by adding another functional symbol which corresponds, roughly speaking, to $x^{\log x^{\log \log x}}$ (see also [50]), then we can easily bound the size of $\binom{n}{k}$, but the problem of defining $\binom{n}{k}$ will still be left open.

We can approach the problem of the definition of $k! = m$ in a different way using the binomial coefficient corresponding to $\binom{2k}{k}$. Recall that,

$$\binom{2k}{k} = \frac{(2k)!}{(k!)^2} = \frac{(k+1)(k+2)\cdots(2k)}{k!}.$$

So we can define

$$k! = m \qquad \text{iff} \qquad m\binom{2k}{k} = (k+1)(k+2)\cdots(2k).$$

Concerning the sizes of the elements involved in the above definition, the following inequalities are true:

$$\binom{2k}{k} < 2^{2k} < m^2 \quad \text{and} \quad (k+1)(k+2)\cdots(2k) \leq \#(m,m).$$

But unfortunately this relation also does not solve all the problems, since there is not an $E_1^\#$-formula, which is known to me, defining $(k+1)(k+2)$ $\cdots(2k)$.

§9. **Residue fields.** In this last section, I will briefly discuss more recent results on residues fields and quotient fields of models of $I\Delta_0 + \Omega_1$. I will omit the proofs which can be found in [16] and in [15].

Pratt showed in [41] that *prime* is in $NP \cap co\text{-}NP$. The difficult part is to show that *prime* $\in NP$. He used Fermat's Little Theorem and the cyclicity of $F_p^*$. Miller proved in [30] that under an Extended Riemann Hypothesis, *prime* is in P. The connection to $I\Delta_0 + \Omega_1$ is the following: Buss in [8] isolated a natural subsystem $S_2^1$ of $I\Delta_0 + \Omega_1$ and showed that any predicate provably in $NP \cap co\text{-}NP$ over $S_2^1$ is in fact in P. This is the motivation for studying Fermat's Little Theorem over $I\Delta_0 + \Omega_1$.

We recall that in a model $\mathcal{M}$ of $I\Delta_0$ the notions of prime and irreducible coincide. Let $p \in \mathcal{M}$, and let $F_p$ be the residue field. For $p$ nonstandard, $F_p$ is a field of characteristic 0. The classical results about finite fields are the following:

(1) $F_p^*$ is cyclic;

(2) $F_p$ has a unique extension $F_{p^n}$ of dimension $n$ for each $n$, which is the splitting field of $x^{p^n} - x$;

(3) the automorphism group of $F_{p^n}$ over $F_p$ is cyclic, generated by $x \mapsto x^p$.

For PA, one knows that for $p$ nonstandard the $F_p$ are up to elementary equivalence the characteristic 0 pseudofinite fields of Ax (see [25]). For IOpen, Macintyre and Marker showed in [26] that if $F_p$ is a field, then it can be any field of characteristic 0 up to elementary equivalence. In particular, a residue field of a model of IOpen may have infinitely many extensions of each dimension.

Let $\mathcal{M}$ be a model of $I\Delta_0 + \Omega_1$ and let $p$ be a prime in $\mathcal{M}$. Any finite extension $K$ of $F_p$ can be interpreted in $\mathcal{M}$, identifying $K$ with the initial segment $p^n$, and the operations of $+$ and $\cdot$ are defined according to the operations of adding and multiplying polynomials. The field $F_p$ and its extensions have characteristic 0. Exponentiation is only a partial function on $\mathcal{M}$. However, there exists a $\Delta_0$-formula in terms of $+, \cdot$, and $\#$ defining the graph of the function

$$\mathcal{M}/n \times \mathcal{M} \longrightarrow \mathcal{M}/n$$
$$(x,y) \mapsto x^y$$

and satisfying the obvious algebraic laws of exponentiation modulo $n$. The totality of such a function is provable in $I\Delta_0 + \Omega_1$. This definition can be extended also to any ring $\Delta_0$-interpretable in $\mathcal{M}$, with the underlying set some $[0, a]$. Moreover, if $K$ is any finite extension of $F_p$, then exponentiation on $K$ behaves well with respect to $\Delta_0$-definable automorphisms; i.e., for any $\sigma \in \text{Aut}(L/K)$ (L finite extension of $K$) $\sigma(x^m) = \sigma(x)^m$ for all $x \in L$ and $m \in \mathcal{M}$.

At the level of $I\Delta_0$, it is not known if such a formula exists. Because of the totality of exponentiation modulo $n$ in $I\Delta_0 + \Omega_1$, we can at least express Fermat's Little Theorem (FLT),

$$\text{if } p \text{ is a prime and } (a, p) = 1 \text{ then } a^{p-1} \equiv 1(\text{mod } p).$$

One of the classical proofs of (FLT) is group-theoretic: the multiplicative group $F_p^*$ is cyclic of order $p - 1$. The crucial step in the proof of this result is that any polynomial of degree $n$ has at most $n$ roots in a field, and this is proved by induction on $n$. There are obstacles in carrying on the inductive argument in $I\Delta_0 + \Omega_1$ since $n$ may be *too big*. Although we need only to show that $x^r - 1$ ($r$ a divisor of $p - 1$) has at most $r$ roots in $F_p$, the inductive step is applied to $x^{r-1} + x^{r-2} + \cdots + x + 1$, which may be too long to be coded.

Fermat's Little Theorem also can be stated as follows: consider the Frobenius map $\sigma: F_p \longrightarrow F_p$ defined as $\sigma(x) = x^p$. Then $\sigma = identity$, i.e., $x^p = x$ for all $x \in F_p$. The inductive argument used in the proof cannot be formalized in $I\Delta_0 + \Omega_1$.

A third proof of (FLT) uses the complete set of residues and functions related to factorial. It seems that there is no $\Delta_0$-definition, even in #, of factorial modulo an integer, for which the totality of the function is provable in $I\Delta_0 + \Omega_1$. It is still an open problem if $I\Delta_0 + \Omega_1 \vdash$ (FLT).

Among the positive results regarding residue fields of a model of $I\Delta_0 + \Omega_1$, there is the classical result on the existence and uniqueness of an extension of each finite degree $n \in \mathbb{N}$. We only sketch the main steps in the proof of this result and we refer to [16] for the details. The proof uses some classical results of Galois theory. First of all, using the weak pigeonhole principle and exponentiation modulo $n$, it can be proved that any element of $F_p^*$ has an order which is less then $2p$. Moreover, among the orders of all elements, there is a maximal one $d$ which is divisible by the orders of all elements. Here $d$ plays the role of $p - 1$, but we do not know if they coincide.

The structure $F_p^*$ is an abelian group and the following decomposition holds. Let $p_i$ be a prime dividing $d$, and

$$A(p_i) = \{x \in F_p^*: \text{order of } x \text{ is a power of } p_i\}.$$

THEOREM 9.1. $F_p^*$ *is the direct product of* $A(p_i)$*s for all* $p_i$*s dividing* $d$*, i.e., each element of* $F_p^*$ *can be written in a unique way as a product of elements each belonging to the* $A(p_i)$*s.*

In fact, for our purposes, we only need to decompose $F_p^*$ as follows:

$$F_p^* = A(r) \times A(r') \qquad (19)$$

where $r$ is a prime of $\mathbb{N}$ dividing $d$ and $A(r')$ contains elements whose order is coprime with $r$. The elements of $A(r')$ are $r$-powers. By purely algebraic arguments, $A(r)$ is cyclic. Let $\alpha$ be a generator of $A(r)$. Then $F_p$ contains the $r$-roots of unity, and these are

$$1, \alpha^{r^{k-1}}, \alpha^{2r^{k-1}}, \dots, \alpha^{(r-1)r^{k-1}}.$$

We can exhibit the coset representative of the $r$-powers of $F_p^*$, and these are

$$1, \alpha, \alpha^2, \dots, \alpha^{r-1}.$$

So, if $r$ divided $d$, then not all elements are $r$-powers, and in order to obtain an extension of $F_p$ of degree $r$, we need only to add an $r$-root of an element which is not an $r$-power.

A similar argument shows the existence of an extension of $F_p$ of degree $n$, for each $n \in \mathbb{N}$ such that $n$ divides $d$. In the hypothesis that $n \in \mathbb{N}$ and $n$ divides $d$, the uniqueness of the normal extension of $F_p$ of degree $n$ follows by a classical result in Kummer theory (see [23]).

Using now some Galois theory, we get a very sharp upper bound on the number of normal extensions of degree $r^k$, for $r$ prime and $r, k \in \mathbb{N}$.

LEMMA 9.2. *For no* $r, k \in \mathbb{N}$, $r$ *prime, does* $F_p$ *have two distinct normal extensions of degree* $r^k$.

These results hold also for any finite extension of $F_p$. Let $K$ be a finite extension of $F_p$. We now describe the Galois group of a finite normal extension of $K$. First we recall the following definition.

DEFINITION 9.3. A group $G$ is called $Z$-group if all its Sylow subgroups are cyclic.

Examples of $Z$-groups are $S_3$ and all the dihedral groups $D_{2n+1}$. $Z$-groups are generated by two elements, and they are solvable, as shown in [40].

THEOREM 9.4. *If* $L$ *is a finite normal extension of* $K$, *then the Galois group of* $L$ *over* $K$ *is a* $Z$-*group*.

The proof of this result is purely algebraic. From Galois theory and properties of $Z$-groups, we get the uniqueness of normal extensions of $F_p$ of each dimension $n \in \mathbb{N}$. In a $Z$-group, being normal for a subgroup is a transitive property. This implies the existence of normal extensions of $F_p$ of dimension $r^k$ for each $k$ and $r$ a prime divisor of $d$.

The remaining case of the existence of a normal extension of degree $r$, when $r$ is a prime not dividing $d$, (i.e., when the field does not contain primitive $r$-roots of unity) is obtained by using some ideas of Albert from [2]. The proof is very technical and it is spelled out in [16]. We only remark here that the proof uses the fact that the $\Delta_0$-definition of exponentiation on $K$ (finite extension of $F_p$) respects $\Delta_0$-automorphisms $\sigma \in \text{Aut}(L/K)$, for any finite extension $L$ of $K$.

We now have the following classical results.

COROLLARY 9.5. *Let $L$ be a normal extension of $K$ of degree $n$. Then the Galois group of $L$ over $K$ is cyclic.*

COROLLARY 9.6. *The norm map $N: L \longrightarrow K$ is surjective for all finite normal extensions $L$ of $K$.*

Models of PA, as well as models of $I\Delta_0$, are not principal ideal domains. If $\mathcal{M}$ is a model of PA and $I$ is an ideal of $\mathcal{M}$, then $I$ is principal if and only if $I$ is definable, and a generator is a prime element. For $\mathcal{M}$ model of $I\Delta_0$, an ideal $I$ is principal if and only if $I$ is $\Delta_0$-definable, and also in this case, a generator is a prime element (recall that in $I\Delta_0$ the notions of prime and irreducible coincide). Let $\mathcal{M}$ be a model of $I\Delta_0$ and $I = \{x \in \mathcal{M}: v_2(x) > \mathbb{N}\}$. It can be easily proved that $I$ is a prime ideal but $I$ is not maximal since $I \subset \langle 2 \rangle$, so $I$ cannot be principal. In [49] Wilkie studied the strength of the axiom,

DIP: any definable ideal is principal,

over the theory of discretely ordered rings. He proved that DOR + DIP is weaker than PA. In [15] axiom DIP is considered over $I\Delta_0$, and it is proved that the theory $I\Delta_0$ + DIP is equivalent to PA. Hence, DOR + DIP $\nvdash I\Delta_0$.

Ideals of models of PA have been considered also in [52].

We now want to extend the result on the existence and uniqueness of a normal extension of each finite degree also to quotient fields $\mathcal{M}/I$, where $\mathcal{M}$ is a model of $I\Delta_0 + \Omega_1$ and $I$ is a maximal ideal of $\mathcal{M}$. The idea is to make the undefinable ideal $I$ of $\mathcal{M}$ into a definable one in a $\Delta_0$-elementary extension $\mathcal{M}^*$ of $\mathcal{M}$. Then we use the fact that for the residue field of $\mathcal{M}^*$ we have already proved the existence and uniqueness of a normal extension of each finite dimension, and that this property can be expressed in a $\Delta_0$-formula. The extension $\mathcal{M}^*$ is obtained via a restricted ultrapower construction.

We work now in $I\Delta_0$. First of all, following some ideas of Cherlin in [9], we establish a correspondence between ideals of $\mathcal{M}$ and filters on the set $\mathcal{D}_P$ of the definable subsets of primes in $\mathcal{M}$. For any $a \in \mathcal{M}$, let $S_a$ be the set of primes dividing $a$.

DEFINITION 9.7. (1) For any ideal $I$ of $\mathcal{M}$, let $\mu(I)$ be the filter generated by the $S_a$s, $a \in I$, i.e., $\mu(I) = \{X \in \mathcal{D}_P: S_{a_1} \cap \cdots \cap S_{a_n} \subseteq X \text{ with } a_1, \ldots, a_n \in I\}$.

(2) If $\mu$ is a filter of $\mathcal{D}_P$, let $I(\mu) = \{a \in \mathcal{M}: S_a \in \mu\}$. A filter of $\mathcal{D}_P$ is bounded if it contains $S_a$, for some $a \in \mathcal{M}$.

The correspondence between $I$ and $\mu$ is stated in the following lemma whose proof is essentially the same as the one of Cherlin in [9]. He uses the prime decomposition of any element which is also available in any model of $I\Delta_0$.

LEMMA 9.8.    (i) *If $I$ is a prime ideal, then $\mu(I)$ is an ultrafilter.*
(ii) *If $\mu$ is a bounded ultrafilter, then $I(\mu)$ is a maximal ideal.*
(iii) *If $I$ is maximal, then $I(\mu(I)) = I$.*
(iv) *If $\mu$ is a bounded ultrafilter, then $\mu(I(\mu)) = \mu$.*
(v) *If $I$ is a proper prime ideal, then $I(\mu(I))$ is the unique maximal ideal containing $I$.*

The restricted ultrapower construction we need is the following. Let $\mathrm{PDef}_0(\mathcal{M}^P)$ be the set of $\Delta_0$-definable functions, from the set of primes $P$ to $\mathcal{M}$, which are bounded by a term-function. (Each term in the language defines a function on a model, and in the case of arithmetic, this is a polynomial function.) If $\mu$ is an ultrafilter of $\mathcal{D}_P$ then we define an equivalence relation on $\mathrm{PDef}_0(\mathcal{M}^P)$ as usual, and we get the term-bounded ultrapower $\mathrm{PDef}_0(\mathcal{M}^P)/\mu$. A version of Łos theorem holds for $\Delta_0$-formulas, and this implies that $\mathcal{M} \prec_{\Delta_0} \mathrm{PDef}_0(\mathcal{M}^P)/\mu$. If $\mathcal{M}$ is a model of $I\Delta_0$ then $\mathrm{PDef}_0(\mathcal{M}^P)/\mu$ is also a model of $I\Delta_0$. This holds since we can express the $\Delta_0$-scheme of induction as follows:

$$\forall \bar{y} \forall t (\theta(0, \bar{y}) \wedge (\forall x \leq t)\theta(x, \bar{y}) \to \theta(x+1, \bar{y})) \to (\forall x \leq t)\theta(x, \bar{y})).$$

REMARK 9.9.  The above results are true also in any expanded language of arithmetic, and, in particular, for $I\Delta_0 + \Omega_1$.

Let $I$ be a nonprincipal maximal ideal of $\mathcal{M}$ model of $I\Delta_0$, and let $\mu = \mu(I)$ be the correspondent ultrafilter. Let $\mathcal{M}^*$ be the term-bounded ultrapower $\mathrm{PDef}_0(\mathcal{M}^P)/\mu$. In $\mathcal{M}^*$, the equivalence class of the identity function on $P$, $id_\mu$, is a prime element, and so the ideal generated by $id_\mu$ is a maximal principal ideal.

THEOREM 9.10.  *In the hypothesis above, $\langle id_\mu \rangle \cap \mathcal{M} = I$ and the function $\eta: \mathcal{M}/I \longrightarrow (\mathrm{PDef}_0(\mathcal{M}^P)/\mu)/\langle id_\mu \rangle$ defined by $\eta(a + I) = a_\mu + \langle id_\mu \rangle$ is an isomorphism.*

So, $I$ is made principal in a $\Delta_0$-elementary extension of $\mathcal{M}$. We do not prove this theorem. We only notice that in order to prove that $\eta$ is surjective, a version of the Chinese Remainder Theorem is used, and this is available in $I\Delta_0$ (see Theorem 3.8). It follows then that $\mathcal{M}/I \equiv (\mathrm{PDef}_0(\mathcal{M}^P)/\mu)/\langle id_\mu \rangle$. Notice that Theorem 9.10 holds also in any extension of the language, so in particular for any model of $I\Delta_0 + \Omega_1$.

Let $\mathcal{M}$ be a model of $I\Delta_0 + \Omega_1$. From the first part of this section, we know that $(\mathrm{PDef}_0(\mathcal{M}^P)/\mu)/\langle id_\mu \rangle$ has a unique normal extension of each finite degree. This property can be expressed by a $\Delta_0$-formula in the language of arithmetic; hence, also $\mathcal{M}/I$ has a unique normal extension of each finite degree.

## REFERENCES

[1] M. AJTAI, *The complexity of the pigeonhole principle*, **Combinatorica. An International Journal on Combinatorics and the Theory of Computing**, vol. 14 (1994), no. 4, pp. 417–433.

[2] A. ADRIAN ALBERT, *Fundamental concepts of higher algebra*, The University of Chicago Press, Chicago, 1958.

[3] ALAN BAKER, *A concise introduction to the theory of numbers*, Cambridge University Press, Cambridge, 1984.

[4] J. H. BENNETT, *On spectra*, Ph.D. thesis, Princeton University, 1962.

[5] ALESSANDRO BERARDUCCI and PAOLA D'AQUINO, $\Delta_0$-*complexity of the relation* $y = \prod_{i \leq n} F(i)$, **Annals of Pure and Applied Logic**, vol. 75 (1995), no. 1-2, pp. 49–56, Proof theory, provability logic, and computation (Berne, 1994).

[6] ALESSANDRO BERARDUCCI and BENEDETTO INTRIGILA, *Combinatorial principles in elementary number theory*, **Annals of Pure and Applied Logic**, vol. 55 (1991), no. 1, pp. 35–50.

[7] ALESSANDRO BERARDUCCI and MARGARITA OTERO, *A recursive nonstandard model of normal open induction*, **The Journal of Symbolic Logic**, vol. 61 (1996), no. 4, pp. 1228–1241.

[8] SAMUEL R. BUSS, *Bounded arithmetic*, Bibliopolis, Naples, 1986.

[9] GREGORY L. CHERLIN, *Ideals of integers in nonstandard number fields*, **Model theory and algebra (A memorial tribute to Abraham Robinson)**, Springer, Berlin, 1975, pp. 60–90. Lecture Notes in Mathematics, Vol. 498.

[10] C. CORNAROS and C. DIMITRACOPOULOS, *The prime number theorem and fragments of* PA, **Archive for Mathematical Logic**, vol. 33 (1994), no. 4, pp. 265–281.

[11] PAOLA D'AQUINO, *Local behaviour of the Chebyshev theorem in models of* $I\Delta_0$, **The Journal of Symbolic Logic**, vol. 57 (1992), no. 1, pp. 12–27.

[12] ——, *A sharpened version of McAloon's theorem on initial segments of models of* $I\Delta_0$, **Annals of Pure and Applied Logic**, vol. 61 (1993), no. 1-2, pp. 49–62, Provability, Interpretability and Arithmetic Symposium (Utrecht, 1991).

[13] ——, *Pell equations and exponentiation in fragments of arithmetic*, **Annals of Pure and Applied Logic**, vol. 77 (1996), no. 1, pp. 1–34.

[14] ——, *Solving Pell equations locally in models of* $I\Delta_0$, **The Journal of Symbolic Logic**, vol. 63 (1998), no. 2, pp. 402–410.

[15] ——, *Quotient fields of a model of* $I\Delta_0 + \Omega_1$, **MLQ Mathematical Logic Quarterly**, vol. 47 (2001), no. 3, pp. 305–314.

[16] PAOLA D'AQUINO and ANGUS MACINTYRE, *Non-standard finite fields over* $I\Delta_0 + \Omega_1$, **Israel Journal of Mathematics**, vol. 117 (2000), pp. 311–333.

[17] C. DIMITRACOPOULOS, *Matijasevič theorem and fragments of arithmetic*, Ph.D. thesis, Manchester University, 1980.

[18] HAIM GAIFMAN and CONSTANTINE DIMITRACOPOULOS, *Fragments of Peano's arithmetic and the MRDP theorem*, **Logic and Algorithmic (Zurich, 1980)**, Univ. Genève, Geneva, 1982, pp. 187–206.

[19] PETR HÁJEK and PAVEL PUDLÁK, *Metamathematics of first-order arithmetic*, Springer-Verlag, Berlin, 1998, Second printing.

[20] G. H. HARDY and E. M. WRIGHT, *An introduction to the theory of numbers*, fifth ed., The Clarendon Press Oxford University Press, New York, 1979.

[21] J. P. JONES and Y. V. MATIJASEVIČ, *Proof of recursive unsolvability of Hilbert's tenth problem*, **The American Mathematical Monthly**, vol. 98 (1991), no. 8, pp. 689–709.

[22] RICHARD KAYE, *Diophantine induction*, **Annals of Pure and Applied Logic**, vol. 46 (1990), no. 1, pp. 1–40.

[23] SERGE LANG, *Algebra*, third ed., Springer-Verlag, New York, 2002.

[24] H. LESSAN, *Models of arithmetic*, Ph.D. thesis, Manchester University, 1978.

[25] ANGUS MACINTYRE, *Residue fields of models of* p, *Logic, methodology and philosophy of science, VI (Hannover, 1979)*, North-Holland, Amsterdam, 1982, pp. 193–206.

[26] ANGUS MACINTYRE and DAVID MARKER, *Primes and their residue rings in models of open induction*, *Annals of Pure and Applied Logic*, vol. 43 (1989), no. 1, pp. 57–77.

[27] KENNETH L. MANDERS and LEONARD ADLEMAN, *NP-complete decision problems for binary quadratics*, *Journal of Computer and System Sciences*, vol. 16 (1978), no. 2, pp. 168–184.

[28] YU. I. MANIN, *A course in mathematical logic*, Springer-Verlag, New York, 1977, Translated from the Russian by Neal Koblitz, Graduate Texts in Mathematics, Vol. 53.

[29] KENNETH MCALOON, *On the complexity of models of arithmetic*, *The Journal of Symbolic Logic*, vol. 47 (1982), no. 2, pp. 403–415.

[30] GARY L. MILLER, *Riemann's hypothesis and tests for primality*, *Seventh annual ACM symposium on theory of computing (Albuquerque, NM, 1975)*, Assoc. Comput. Mach., New York, 1975, pp. 234–239.

[31] M. OTERO, *Models of open induction*, Ph.D. thesis, Oxford University, 1991.

[32] MARGARITA OTERO, *The amalgamation property in normal open induction*, *Notre Dame Journal of Formal Logic*, vol. 34 (1993), no. 1, pp. 50–55.

[33] ———, *The joint embedding property in normal open induction*, *Annals of Pure and Applied Logic*, vol. 60 (1993), no. 3, pp. 275–290.

[34] ———, *Quadratic forms in normal open induction*, *The Journal of Symbolic Logic*, vol. 58 (1993), no. 2, pp. 456–476.

[35] ROHIT PARIKH, *Existence and feasibility in arithmetic*, *The Journal of Symbolic Logic*, vol. 36 (1971), pp. 494–508.

[36] J. PARIS and A. WILKIE, *Counting $\Delta_0$ sets*, *Polska Akademia Nauk. Fundamenta Mathematicae*, vol. 127 (1987), no. 1, pp. 67–76.

[37] J. B. PARIS, A. J. WILKIE, and A. R. WOODS, *Provability of the pigeonhole principle and the existence of infinitely many primes*, *The Journal of Symbolic Logic*, vol. 53 (1988), no. 4, pp. 1235–1244.

[38] JEFF PARIS and LEO HARRINGTON, *Mathematical incompleteness in Peano Arithmetic*, *Handbook of mathematical logic*, North-Holland Publishing Co., Amsterdam, 1977, pp. 1133–1142.

[39] JEFF B. PARIS and CONSTANTINE DIMITRACOPOULOS, *Truth definitions for $\Delta_0$ formulae*, *Logic and algorithmic (Zurich, 1980)*, Univ. Genève, Geneva, 1982, pp. 317–329.

[40] DONALD PASSMAN, *Permutation groups*, W. A. Benjamin, Inc., New York-Amsterdam, 1968.

[41] VAUGHAN R. PRATT, *Every prime has a succinct certificate*, *SIAM Journal on Computing*, vol. 4 (1975), no. 3, pp. 214–220.

[42] PAVEL PUDLÁK, *A definition of exponentiation by a bounded arithmetical formula*, *Commentationes Mathematicae Universitatis Carolinae*, vol. 24 (1983), no. 4, pp. 667–671.

[43] J. C. SHEPHERDSON, *A non-standard model for a free variable fragment of number theory*, *Bull. Acad. Polon. Sci. Sér. Sci. Math. Astronom. Phys.*, vol. 12 (1964), pp. 79–86.

[44] RICHARD SOMMER, *Ordinal arithmetic in $I\Delta_0$*, *Arithmetic, proof theory, and computational complexity (Prague, 1991)*, Oxford Univ. Press, New York, 1993, pp. 320–363.

[45] GAISI TAKEUTI, *Two applications of logic to mathematics*, Iwanami Shoten, Publishers, Tokyo, 1978, Kanô Memorial Lectures, Vol. 3, Publications of the Mathematical Society of Japan, No. 13.

[46] LOU VAN DEN DRIES, *Some model theory and number theory for models of weak systems of arithmetic*, *Model theory of algebra and arithmetic (Proc. Conf., Karpacz, 1979)*, Springer, Berlin, 1980, pp. 346–362.

[47] A. WILKIE and J. PARIS, *On the existence of end extensions of models of bounded induction*, *Logic, methodology and philosophy of science, VIII (Moscow, 1987)*, North-Holland, Amsterdam, 1989, pp. 143–161.

[48] A. J. WILKIE, *Some results and problems on weak systems of arithmetic*, **Logic Colloquium '77 (Proc. Conf., Wrocław, 1977)**, North-Holland, Amsterdam, 1978, pp. 285–296.

[49] ———, *On discretely ordered rings in which every definable ideal is principal*, **Model theory and arithmetic (Paris, 1979–1980)**, Springer, Berlin, 1981, pp. 297–303.

[50] A. J. WILKIE and J. B. PARIS, *On the scheme of induction for bounded arithmetic formulas*, **Annals of Pure and Applied Logic**, vol. 35 (1987), no. 3, pp. 261–302.

[51] A. WOODS, **Some problems in logic and number theory and their connections**, Ph.D. thesis, Manchester University, 1981.

[52] A. R. WOODS, **Algebraic properties of non-standard numbers**, Master's thesis, Monash University, 1977.

DIPARTIMENTO DI MATEMATICA
SECONDA UNIVERSITA' DI NAPOLI
VIA VIVALDI, 43
CASERTA, ITALY
*E-mail*: paola.daquino@unina2.it

# Index for *Weak fragments of Peano Arithmetic*

# LECTURE NOTES IN LOGIC
## General Remarks

This series is intended to serve researchers, teachers, and students in the field of symbolic logic, broadly interpreted. The aim of the series is to bring publications to the logic community with the least possible delay and to provide rapid dissemination of the latest research. Scientific quality is the overriding criterion by which submissions are evaluated.

Books in the Lecture Notes in Logic series are printed by photo-offset from master copy prepared using LaTeX and the ASL style files. For this purpose the Association for Symbolic Logic provides technical instructions to authors. Careful preparation of manuscripts will help keep production time short, reduce costs, and ensure quality of appearance of the finished book. Authors receive 50 free copies of their book. No royalty is paid on LNL volumes.

Commitment to publish may be made by letter of intent rather than by signing a formal contract, at the discretion of the ASL Publisher. The Association for Symbolic Logic secures the copyright for each volume.

The editors prefer email contact and encourage electronic submissions.

## Editorial Board

Samuel R. Buss, Managing Editor
Department of Mathematics
University of California, San Diego
La Jolla, California 92093-0112
sbuss@ucsd.edu

Shaughan Lavine
Department of Philosophy
The University of Arizona
P.O. Box 210027
Tuscon, Arizona 85721-0027
shaughan@ns.arizona.edu

Anand Pillay
Department of Mathematics
University of Illinois
1409 West Green Street
Urbana, Illinois 61801
pillay@math.uiuc.edu

Lance Fortnow
Department of Computer Science
University of Chicago
1100 East 58th Street
Chicago, Illinois 60637
fortnow@cs.uchicago.edu

Steffen Lempp
Department of Mathematics
University of Wisconsin
480 Lincoln Avenue
Madison, Wisconsin 53706-1388
lempp@math.wisc.edu

W. Hugh Woodin
Department of Mathematics
University of California, Berkeley
Berkeley, California 94720
woodin@math.berkeley.edu

# Editorial Policy

1. Submissions are invited in the following categories:
i) Research monographs                iii) Reports of meetings
ii) Lecture and seminar notes          iv) Texts which are out of print
Those considering a project which might be suitable for the series are strongly advised to contact the publisher or the series editors at an early stage.

   2. Categories i) and ii). These categories will be emphasized by Lecture Notes in Logic and are normally reserved for works written by one or two authors. The goal is to report new developments quickly, informally, and in a way that will make them accessible to non-specialists. Books in these categories should include
– at least 100 pages of text;
– a table of contents and a subject index;
– an informative introduction, perhaps with some historical remarks, which should be accessible to readers unfamiliar with the topic treated;

   In the evaluation of submissions, timeliness of the work is an important criterion. Texts should be well-rounded and reasonably self-contained. In most cases the work will contain results of others as well as those of the authors. In each case, the author(s) should provide sufficient motivation, examples, and applications. Ph.D. theses will be suitable for this series only when they are of exceptional interest and of high expository quality.

   Proposals in these categories should be submitted (preferably in duplicate) to one of the series editors, and will be refereed. A provisional judgment on the acceptability of a project can be based on partial information about the work: a first draft, or a detailed outline describing the contents of each chapter, the estimated length, a bibliography, and one or two sample chapters. A final decision whether to accept will rest on an evaluation of the completed work.

   3. Category iii). Reports of meetings will be considered for publication provided that they are of lasting interest. In exceptional cases, other multi-authored volumes may be considered in this category. One or more expert participant(s) will act as the scientific editor(s) of the volume. They select the papers which are suitable for inclusion and have them individually refereed as for a journal. Organizers should contact the Managing Editor of Lecture Notes in Logic in the early planning stages.

   4. Category iv). This category provides an avenue to provide out-of-print books that are still in demand to a new generation of logicians.

   5. Format. Works in English are preferred. After the manuscript is accepted in its final form, an electronic copy in LaTeX format will be appreciated and will advance considerably the publication date of the book. Authors are strongly urged to seek typesetting instructions from the Association for Symbolic Logic at an early stage of manuscript preparation.

# LECTURE NOTES IN LOGIC

From 1993 to 1999 this series was published under an agreement between the Association for Symbolic Logic and Springer-Verlag. Since 1999 the ASL is Publisher and A K Peters, Ltd. is Co-publisher. The ASL is committed to keeping all books in the series in print.

Current information may be found at http://www.aslonline.org, the ASL Web site. Editorial and submission policies and the list of Editors may also be found above.

Previously published books in the *Lecture Notes in Logic* are:

1. *Recursion theory.* J. R. Shoenfield. (1993, reprinted 2001; 84 pp.)

2. *Logic Colloquium '90; Proceedings of the Annual European Summer Meeting of the Association for Symbolic Logic, held in Helsinki, Finland, July 15–22, 1990.* Eds. J. Oikkonen and J. Väänänen. (1993, reprinted 2001; 305 pp.)

3. *Fine structure and iteration trees.* W. Mitchell and J. Steel. (1994; 130 pp.)

4. *Descriptive set theory and forcing: how to prove theorems about Borel sets the hard way.* A. W. Miller. (1995; 130 pp.)

5. *Model theory of fields.* D. Marker, M. Messmer, and A. Pillay. (1996; 154 pp.)

6. *Gödel '96; Logical foundations of mathematics, computer science and physics; Kurt Gödel's legacy. Brno, Czech Republic, August 1996, Proceedings.* Ed. P. Hajek. (1996, reprinted 2001; 322 pp.)

7. *A general algebraic semantics for sentential objects.* J. M. Font and R. Jansana. (1996; 135 pp.)

8. *The core model iterability problem.* J. Steel. (1997; 112 pp.)

9. *Bounded variable logics and counting.* M. Otto. (1997; 183 pp.)

10. *Aspects of incompleteness.* P. Lindstrom. (1997, 2nd ed. 2003; 163 pp.)

11. *Logic Colloquium '95; Proceedings of the Annual European Summer Meeting of the Association for Symbolic Logic, held in Haifa, Israel, August 9–18, 1995.* Eds. J. A. Makowsky and E. V. Ravve. (1998; 364 pp.)

12. *Logic Colloquium '96; Proceedings of the Colloquium held in San Sebastian, Spain, July 9–15, 1996.* Eds. J. M. Larrazabal, D. Lascar, and G. Mints. (1998; 268 pp.)

13. *Logic Colloquium '98; Proceedings of the Annual European Summer Meeting of the Association for Symbolic Logic, held in Prague, Czech Republic, August 9–15, 1998.* Eds. S. R. Buss, P. Hájek, and P. Pudlák. (2000; 541 pp.)

14. *Model Theory of Stochastic Processes.* S. Fajardo and H. J. Keisler. (2002; 136 pp.)

15. *Reflections on the Foundations of Mathematics; Essays in honor of Solomon Feferman.* Eds. W. Seig, R. Sommer, and C. Talcott. (2002; 444 pp.)

16. *Inexhaustibility; a non-exhaustive treatment.* T. Franzén. (2004; 255 pp.)

17. *Logic Colloquium '99; Proceedings of the Annual European Summer Meeting of the Association for Symbolic Logic, held in Utrecht, Netherlands, August 1–6, 1999.* Eds. J. van Eijck, V. van Oostrom, and A. Visser. (2004; 208 pp.)

18. *The Notre Dame Lectures.* Ed. P. Cholak. (2005, 185 pp.)

19. *Logic Colloquium 2000; Proceedings of the Annual European Summer Meeting of the Association for Symbolic Logic, held in Paris, France, July 23–31, 2000.* Eds. R. Cori, A. Razborov, S. Todorčević, and C. Wood. (2005; 408 pp.)

20. *Logic Colloquium '01; Proceedings of the Annual European Summer Meeting of the Association for Symbolic Logic, held in Vienna, Austria, August 1–6, 2001.* Eds. M. Baaz, S. Friedman, and J. Krajíček. (2005, 486 pp.)

Printed and bound by CPI Group (UK) Ltd, Croydon, CR0 4YY

26/10/2024

01779666-0001